亚太名家荟萃

豪宅赏析

ASIA-PACIFIC MASTERS COLLECTION
MANSION APPRECIATION

席卷港台 SWEEPING HONGKONG AND TAIWAN
唐忠汉　戴勇　许天贵　李文心　凌子达　李益中　刘卫军　蔡馥韩　孟也　高文安　连自成　葛亚曦

风靡大陆 SWEEPING MAINLAND
郑树芬　谭精忠　张清平　黄国桓　罗海峰　傅琼慧　桂峥嵘　王小根　张成喆　江欣宜

全筑装饰　刘荣禄

深圳视界文化传播有限公司 编

中国林业出版社
China Forestry Publishing House

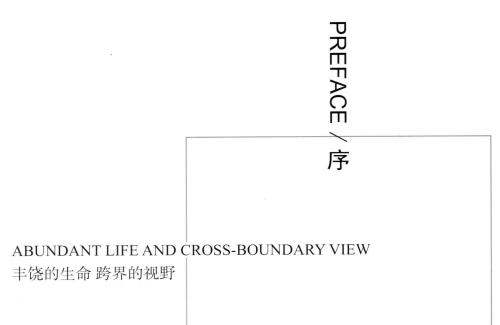

PREFACE / 序

ABUNDANT LIFE AND CROSS-BOUNDARY VIEW
丰饶的生命 跨界的视野

Keeping to your original aspiration leads you to success. This is the appropriate summarization of design. Every designer should keep the original aspiration in mind and maintain vocational self-confidence and professional spirit.

An excellent interior designer must love life, and only in this way can he make a good design. At the same time, you need to improve your designs with the change of time, space and experience. For example, I used to like simple and comfortable design, then my concepts changed a lot, and now I think the more valuable overall judgment is paying more attention on soft decoration to change the temperament of the house and less on hard decoration. In some degrees, soft decoration is a present of personality and thus more important. Sometimes it is decorated by instinct with big rules and without small rules. How to have a more personal and distinct decoration needs to be good at receiving new things. And this is a necessary gift for a designer.

In the design road for so many years, the greatest pleasure is I can constantly learn. I myself was not a professional interior designer and engaged in a variety of industries, such as foreign trade, hotel management and financial management. In the cross-boundary and diversified society nowadays, the so-called professional counterpart is less and less taken for granted in the previous planned economy era. Being active to be involved in various fields to form a cross-boundary view is good for the profession. I have a tip for designers and young men who are interested in being designers, that is learning whatever you like. I suggest G&K designers to broaden international horizons through cross-country travels, learn English and drive. Constantly to learn survival skills and professional knowledge to improve ourselves, let's accomplish our common career.

Nowadays many designers pay attention to self style and individuation, what I want to emphasize is that the design elements should not be too dogmatic and beautiful things are beautiful at any time. Beautiful things from thousands of years ago are still very beautiful. When we look back to Greek and Roman, we'll find that elements from classic world are lasting and charming after a long time with vigorous artistic vitality. In this sense, I am opposed to separate classic, modern and Chinese style. On the one hand, we should respect the original beauty and the law of itself, on the other hand, we can't be limited by inherent elements.

In addition, a designer is not an artist who is very independent and pretty ego and absorbed in the own art world temporarily ignoring the audiences. However for a designer, understanding the clients and communicating with them are very important. This means a designer should have all kinds of cultivations and can present the hazy idea of the clients through professional understandings. But it is not simply catering to the clients. If we meet clients who stick to their different ideas which are not optimistic to implement from our professional judgments, we need to guide them with professional views. This is true for the sake of them. And now their professional abilities and views are gradually increased, more and more people expect to get higher quality design services. So the future development trend will be more optimistic. In such an environment, we need to seek the cause in ourselves and constantly improve and perfect ourselves. As the tile puts, "abundant life and cross-boundary view" are the essences for designers to design.

In the future, an irresistible trend is that more personalized designs will be more popular in the market and designers who are better at communication will be more acceptable by the clients. We must conform to this trend.

G&K Design Grace Kwai

不忘初心，方得始终！用这句话来概括设计之路是再恰当不过的。每一个设计师应该始终把这份初心放在心上，保持职业自信和专业精神。

一位优秀的室内设计师必定是一个热爱生活的人，只有这样才能做出好的设计。同时，对于设计你必须随着时间、空间、阅历的改变而有所增长。比如我以前喜欢比较简约而自在的设计，但后来我的观念发生了一定的变化，现在的我觉得更有价值的整体判断是轻硬装重软装，通过软装来改变房子的气质。软装布置在某种意义上更是个性的张扬，从而也更重要，软装布置很多时候是靠直觉，有大规律没有小规律。如何在空间的装饰上更有个性，与众不同，就要善于接受新事物，这也是一个设计师应该具备的禀赋。

在设计这条路上这么多年，最大的乐趣就是可以不断地学。我自身不是室内设计专业出身，之前从事过多种行业，比如对外贸易、酒店管理和财务管理，在跨界和多元的当今，原来计划经济时代我们讲究的所谓专业对口越来越显得不是那么理所当然。采取主动策略涉及各种领域，形成跨界的视野，对专业有很多帮助。我给设计师和有志于从事设计职业的年轻人有一个忠告：只要喜欢的事情就去学习。比如我建议G&K设计师通过跨国旅行开拓国际化的视野，学英语和驾驶等。在生存技能和专业知识上不断学习，完善自己，完成我们共同的事业。

当下很多设计师讲究自我风格和个性化，我想强调的是设计里的元素不要太教条，美的东西无论在什么时候都是美的。几千年前美的东西现在看来依然很美，当我们回首希腊、罗马，我们会发现古典世界的元素历久弥新，仍然具有旺盛的艺术生命力。从这个意义上，我反对将古典、现代、中式割裂的做法。一方面，我们要尊重美的原汁原味和本身的规律，另一方面，我们不能被固有的元素局限住自己的思路。

此外，设计师不是艺术家，艺术家是很特立独行、相当自我的，很多时候可以暂时忽略受众沉浸在自己的艺术世界中，而对设计师来说，理解客户和与客户沟通的能力显得极为重要。这就是说，设计师要有各种修养，要把甲方的可能还比较朦胧的想法通过专业的理解呈现出来。但这并不是简单的迎合客户，如果遇到坚持自己不同想法的甲方，而我们设计师从专业角度判断其想法执行起来并不乐观，我们设计师就要用专业的观点去引导，这才是真正为客户着想。何况现在业主的专业能力和视野都在逐渐提升，越来越多的人希望得到更高品质的设计服务，未来的发展趋势更加乐观。在这样一种大环境下，我们设计师需要做的就是反求诸己、不断提升和完善自己。正如标题告诉我们的"丰饶的生命，跨界的视野"是设计师安身立命的根本。

在未来，一个不可阻挡的潮流就是，越是个性化的设计越是受市场欢迎的设计，越是善于沟通的设计师越是被业主悦纳的设计师，我们必须顺应这个潮流。

上海桂睿诗建筑设计咨询有限公司 桂峥嵘

056 –081
082 –091 
112 –119 
010 –055
trendzône / DecorGroup
全 築 装 饰
158 –177
092 –103
104 –111
188 –197 
128 –137
198 –209
120 –127
138 –157 
178 –187

CONTENTS / 目录

风靡大陆
SWEEPING MAINLAND

010 | 一揽泰式原生风情 ENJOYING ORIGINAL THAI-STYLE

036 | 写意博览成大家 EXPRESSING THE MIND AND EXPERIENCING THE WORLD CONTRIBUTE TO A MASTER

056 | 传承中的形与意 INHERITANCE OF FORM AND MEANING

072 | 让生活向艺术再进一步 MAKE LIFE ONE STEP CLOSER TO ART

082 | 豪宅匠心 独具风韵 MANSION WITH UNIQUE CHARM

092 | 典雅空间 奢华风尚 ELEGANT SPACE, LUXURIOUS STYLE

104 | 大艺术家·厢 THE ARTIST AND WING

112 | 和谐惬意家 A HARMONIOUS AND COZY HOME

120 | 人和房子，需要慢慢培养感情 HUMAN AND HOUSE NEED TO CULTIVATE RELATIONSHIPS SLOWLY

128 | 静享悠然生活 QUIETLY ENJOYING LEISURE LIFE

138 | 塞纳河畔的柔情 TENDERNESS OF THE SEINE RIVERSIDE

158 | 天人合一 精神内敛 ONENESS OF MAN AND NATURE, RESTRAINED SPIRITS

178 | 天青处 THE AZURE PLACE

188 | 少即是多 LESS IS MORE

198 | 海派密林里的新中式 A NEO-CHINESE STYLE IN VARIOUS SHANGHAI STYLES

席卷港台
SWEEPING
HONGKONG
AND
TAIWAN

212 | 艺玩·质朴 ART INNOVATION AND PURELY INSPIRED

224 | 雅奢主张：文化无界 ADVOCATING ELEGANT LUXURY, CULTURE WITHOUT BOUNDARY

236 | 中体西用的新东方艺术 NEW ORIENTAL ART OF WESTERNIZED CHINESE STYLE

252 | 几何的深度 DEPTH OF GEOMETRY

264 | 大景 轴线 THE VENUE WITH A VIEW, A TALE OF TWO AXES

274 | 质域 TEXTURES

282 | 精锐音悦厅 EXQUISITE AND ENJOYABLE MUSIC HALL

290 | 享·品味 ENJOYMENT AND TASTE

298 | 客制化奢华风 内敛气质别墅 CUSTOMIZED LUXURIOUS STYLE, RESTRAINED CHARMING VILLA

310 | 优享透心美宅 ENJOYING THE BEAUTIFUL RESIDENCE

风靡大陆
SWEEPING MAINLAND

高文安 / Kenneth Ko

一揽泰式原生风情
Enjoying original Thai-style
010

写意博览成大家
Expressing the mind and experiencing the world contribute to a master
036

香港、深圳高文安设计有限公司 创始人

高文安，一九四三年生，香港资深高级室内设计师、香港建筑师学院院士、英国皇家建筑师学院院士、澳洲皇家建筑师学院院士。逾四十年的设计生涯里，完成室内设计五千余项，出版系列作品集《品鉴·品味》《品鉴·传奇》《品鉴·悟道》等，被誉为"香港室内设计之父"。

1976年创办香港高文安设计有限公司，2003年创办深圳高文安设计有限公司，2006年在上海一九三三老场坊、成都宽窄巷子成立设计分公司。2007年创办深圳高文安企业管理有限公司，自创MY系列九大生活品牌。2013年获香港室内设计协会终身成就奖；2014年，获IFI国际室内建筑师设计师联合会"重大国际成就表彰"。2015年，与上市公司宝鹰股份结成战略同盟，出任宝鹰集团副总裁，同年入选《福布斯》中文版"中国最具影响力设计师三十强"。

一揽泰式原生风情
ENJOYING ORIGINAL THAI-STYLE

项目名称 | 华发水郡花园三期A区W4别墅
设计公司 | 深圳高文安设计有限公司
设 计 师 | 高文安
项目地点 | 广东珠海
项目面积 | 800 m²
摄 影 师 | KKD推广部
主要材料 | 云石、文化石、实木、锈砖、木地板、布帘、竹帘、纱帘等

DESIGN CONCEPT | 设计理念

This project is located in Clear Lake waterside of the provincial wetland park. Inspired by The Silk Road at sea, KKD uses design to step over seas and oceans to endow the wetland villa with Southeast Asian island civilization and exquisite tastes, which creates an original romantic living space and a lifestyle from vanity to simplicity and from noise to tranquility, giving busy urbanites absolute relaxing home atmospheres to relieve and release their bodies and minds.

　　本案位于省级湿地公园清湖水畔，KKD以海上丝绸之路为灵感，用设计跨海越洋，将东南亚特色岛屿文明及精致品味移植到湿地别墅，打造原生态的浪漫居家生活空间，创造一种从浮华走向平实、从喧闹回归宁静的生活方式，让繁忙都市人在绝对放松的家庭氛围里，获得身心的舒缓和释放。

Living room and dining room, the breath of forest
客餐厅·森林的呼吸

Home design is actually life design. Southeast Asia is located in rainy and fertile tropical zone where the biggest feature of home life is nature with distinct tropical rainforest style. As for the wetland villa, Kenneth Ko starts from the narrow place and transforms the ten steps alley into an avenue covered with black cobblestone and white marbles. The lush tropical forest is decorated with South Asia floor lamps, which is natural and romantic. Walking around, the breath can be fresh with greenery.

Opening the rosewood doors, the seemingly casual displays, such as Buddha head, wood carving, pottery and woven carpet reflect the original and perceptual Southeast Asian style.

Entering the living room, a whole glass feature wall makes the simple and elegant Thai-style space bathed in natural lights. The grasses are green and the trees are lush outside the window, which integrates with the insides into one. Log is the soul of the living room decoration. From the overlapping ceiling transformed from Southeast Asian slope roof, the hollow folding screen to log floor and tables and chairs, each is a deduction of Southeast Asian log.

Southeast Asian home decoration is full of relaxing and comfortable holiday tone. The living room uses furniture with concise lines and elegant colors to create a fresh and comfortable atmosphere, collocating with gold Buddha and phoenix sculpture, which manifests the holy and perceptual features of Thai-style.

The decorative fireplace made of natural marble with natural surfaces and the hanging Thai-style Buddha spire copper crafts have some original religious meanings in the comfortable vision, which is reassuring.

The dining room continues the taste of returning to nature and pursuing originality as in the living room. The panoramic French window and the seasonal changed garden landscape are beautiful enough to feast the eyes. Details such as romantic and elegant lotus droplight, wood grid ceiling and Southeast Asian floral wood furniture with Chinese elements are worth to taste.

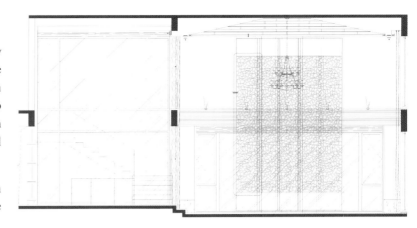

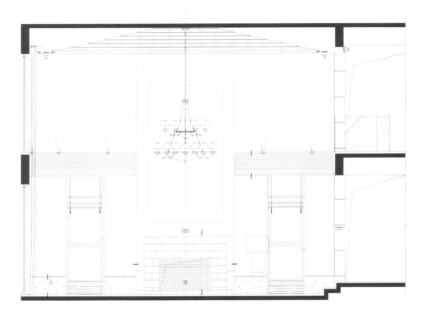

家居设计实质上是对生活的设计，东南亚地处多雨富饶的热带，家居生活最大特点便是崇尚自然，别具的热带雨林风情大行其道。湿地别墅，高文安于窄处着笔，将长不过十步的小巷，设计成黑色鹅卵石与白色大理石板铺就的林荫道，繁茂生长的热带林木间，点缀南亚风情的落地灯，自然、浪漫，漫步其中，呼吸都带着绿意。

推开红木大门，佛头、木雕、陶罐、编织地毯，看似漫不经心的陈设，东南亚原始感性的风情初露端倪。

步入客厅，一整面玻璃景观墙，让简丽的泰式空间沐浴自然光照，窗外芳草依依、绿树成荫，室内与户外真正做到浑然一体。原木构成客厅装饰的灵魂，上至东南亚坡屋顶形式演变的叠级天花，下至镂空结构的折叠屏风、原木地板与桌椅，无一不是东南亚的木作演绎。

东南亚风格家居，充满轻松自在的假日情调，客厅以线条简洁、颜色素雅的家具营造清爽舒适的氛围，搭配金身佛像、凤凰鸟雕塑装饰有度，泰式神圣而感性的特色尽显淋漓。

以自然面的天然大理石砌成的装饰性壁炉，悬挂上泰式佛塔尖顶铜艺品，舒服的视觉感里包含几分原始的虔诚意味，叫人心安。

餐厅很好延续了客厅回归自然、追求原汁原味的品味。全景落地窗，园林景观四季变换，秀色可餐。浪漫雅趣的莲台吊灯、木格天花，以及融入了中式元素的东南亚特色雕花木家具，每一个细节都值得品味。

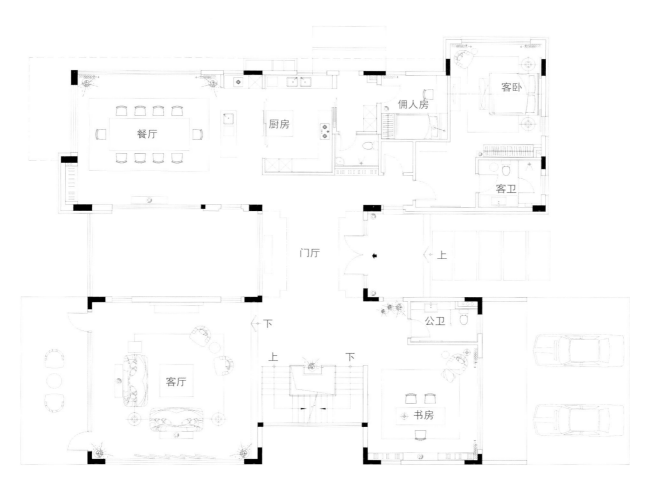

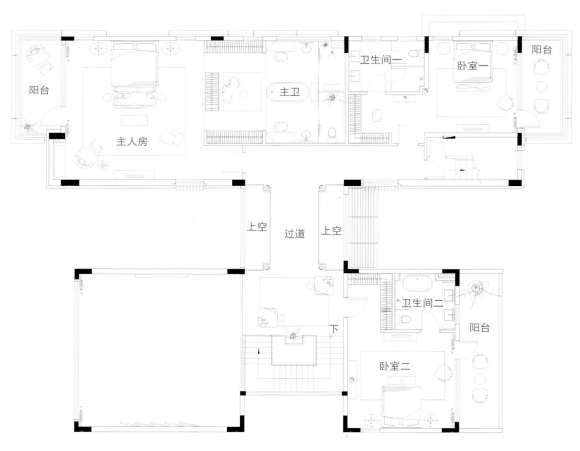

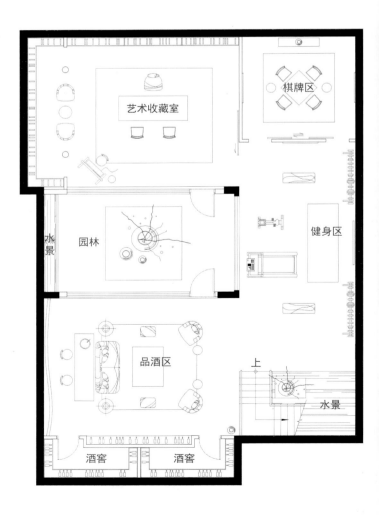

Bedroom, original Thai-style scenery
卧室·泰乡原风景

Kenneth Ko has poured Thai-style into the design of bedroom. The half-open space and the dark and light colors bring out Thai-style elegance and dignity. Log furniture, Thai-style floral background wall and charming yarn curtain, every decorative element reveals Southeast Asian amorous feelings, collocating with Chinese hand painted screen, which creates an elegant living space with tropical flavor and modern sense.

In Thailand with sufficient sunshine, the rhythmic lights and shadows in primeval forest are full of natural poetry. In the elder's room, KKD team uses splendid sunshine and elaborately designed lights to create an aesthetic change of light and shadow. Retro floral bedstead collocates with gold flower carpet, which is sedate and noble.

In the guest bedroom, the pea green linen wallpaper and carpet, dying cotton quilt in lotus leaf shape match with Thai-style cloth and copper wall furnishings, integrating modern with classical, which is fresh and elegant.

In the guest room, log shutter, cool and fresh cane chair, Thai silk pillow and boat foot rest constitute invariable Thai feelings. Being here is like being in the romantic country with an ancient legend. The distant local conditions and customs in Thailand are elegant, leisure and full of Zen-like living taste which are easy to get.

卧室的设计上，高文安倾注了泰国情怀。半开放式空间，深浅交错的色调，带出泰式高雅与稳重。原木家私、泰式雕花背景墙、妩媚纱缦，每一个装饰元素都透出南洋风情，搭配中式手绘屏风，营造出热带气息与现代感兼备的优雅生活空间。

日照充足的泰乡，原始森林律动的光影充满天然的诗意。老人房，KKD团队借用灿烂的阳光与精心设计的灯光，营造出唯美的光影变化。复古的雕花床架，搭配金色的编花地毯，沉稳中显露贵气。

客卧，豆绿色的麻质墙纸地毯，荷叶图纹的印染棉被，点缀泰式布艺与铜艺墙面装饰品，穿插现代与古典，清新雅致。

客房，原木百叶窗、清凉藤椅、泰丝抱枕、船形脚踏，构成不变的泰乡情怀。置身其中，好似亲临那个有着古老传说的浪漫国度，遥远泰乡的风土人情，清雅、休闲又充满禅味的生活情趣触手可及。

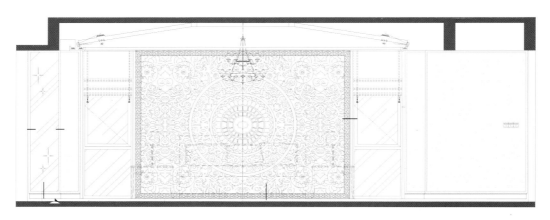

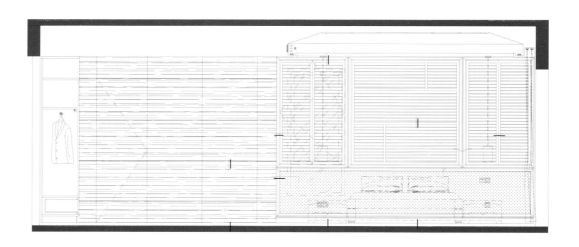

亚太名家荟萃·风靡大陆 | 031

Recreational area, alliance of wood and stone
休闲区·木石之盟

The applications of materials of Southeast Asian home decoration are quaint and unique. Stone and wood couple hardness with softness. In the study and wine area, solid wood furniture, Buddha head with heavy religious complex and distinctive carpet and lamps integrate nobility with nationality, which reflects tranquil Zen and casual life philosophy.

Cultures of Southeast Asian countries, especially in Thailand, are influenced by Indian Buddhism and Brahmanism with strong religious feelings. Corridor droplight inspires by Thailand Loi Krathong Festival which integrates Buddhism and Brahmanism. The globular droplight hangs in the wood hollow ceiling, ups and downs, as if flying sky lanterns, which is like a fancy dream. The colorful carpet makes the wood and stone themed space passionate and romantic.

Ascending upstairs along the corridor to the rooftop, four wooden crafts hanging on the wall are like Konoha butterflies in the dance group, relating late autumn whispers of mountains and forests.

Looking faraway, the lake scenery is very beautiful and wonderful. Here you can enjoy the poetic amorous feelings of living near the lake.

东南亚家居，材料的运用上古雅独到，石与木，刚柔并济。书房与品酒区，实木家私，带有浓郁宗教情结的佛头，以及别具特色的地毯、灯具，高贵和民族融为一体的格调，投射出静谧的禅味，以及随性生活的哲理。

南洋诸国，特别是泰国，其文化受到印度佛教与婆罗门教的渗透，带有浓烈的宗教色彩。楼道吊灯，设计灵感来源于融合佛教盂兰盆会的泰国水灯节，悬挂于木结构镂空天花的球状吊灯，上下起伏错落，似飘飞的天灯，如梦似幻。辅以色彩鲜艳的地毯装点，让木石主题的空间洋溢澎湃的热情与浪漫。

顺着楼道拾级而上，登上楼顶天台，墙上挂着的四幅木雕工艺品，像团舞的木叶蝶，述说深秋山林的秋日私语。

凭栏远望，湖光潋滟晴方好，山色空蒙雨亦奇，在水天一色波光里，体验临湖而居的诗意风情。

写意博览成大家
EXPRESSING THE MIND AND EXPERIENCING THE WORLD CONTRIBUTE TO A MASTER

项目名称｜华侨城LOFT A座1单元19A
设计公司｜深圳高文安设计有限公司
设 计 师｜高文安
项目地点｜广东深圳
项目面积｜320 m²
摄 影 师｜KKD推广部
主要材料｜石材、实木、清镜、钢板、玻璃、枕木、乳胶漆等

DESIGN CONCEPT ｜ 设计理念

Home has a very special meaning to Kenneth Ko. These "homes" around the world are more gifts which he sends to himself and the world than elaborate designs which he spares efforts to. They carry his life experiences and depositions in different ages and can comfort his soul in long years.

　　家，对于高文安有着极为特殊的意义，这些遍布世界各地的"家"，与其说是他投入时间心血的精心设计，不如说是他送给自己，送给世界的礼物。其承载了高文安不同年龄的人生历练与沉淀，也在漫漫岁月长河里反馈与他心灵的慰藉。

Expressing the mind
写意原心

Openness is always the design concept adhered by Kenneth Ko. Without exception, he transforms thoroughly the entire interior space this time and breaks the spatial separation. The living room, dining room, kitchen and gym are linked into a whole. The transparent glass wall and mirror bring in natural lights. The interior echoes with the outside. The natural changes and drops of lights and shadows are enjoyable. It makes outside scenery fully available to view. The skillful designs make the whole space transparent and spacious, which reflects the calmness and generosity of the owner.

The ubiquitous "wood" is almost used in every space, such as the special wood floor and rough primitive retro wood cabinet whose design manuscripts have changed ten times, entire log headboard of the master bedroom, long crosstie device of the balcony and even the door handle which is specially made of a 5 centimeters solid wood. As for the reasons, one is that Kenneth Ko loves the natural, simple and straightforward wood, and the other is that wood is enduring and as common in both Chinese and Western cultures as the simplest food, which is simple yet tasty. It can be masculine and warm, which depends on how to "cook".

开放性一直是高文安坚持的设计主张，不例外，此次他同样对内部空间进行了贯通改造，打破空间隔阂。将客厅、餐厅、厨房、健身房连为一体，通透的玻璃幕墙与镜面将天然光尽收其中、内外呼应，光影的自然变化和洒落已足够惹味，更令室外风景一览无余。如此巧妙运用，让整个空间通透敞怀，有一种置身无垠的开阔，更见主人的从容豁达。

随处可见的"木"，几乎遍布每一个空间，从单是设计手稿就改了十次的特制木地板、粗糙原始的复古木柜，到主卧床头的整块原木床头板、阳台的长条枕木装置，甚至连大门把手都是由一块5公分厚的实木面特别改造而成。究其缘由，一是高文安喜爱木头的天然朴直，二是木经久不衰，在中西文化中都相当常见，如同最质朴无华的食材，最简单却最易出味。可阳刚可温暖，全看如何"烹饪"。

Experiencing the world
博览众生

The oil painting *Temptation of ST Antony Part* by Australian artist Noel Tunks in the living room tells the story of British queen Elizabeth I. Kenneth Ko loves it very much. Since he bought it from Australia in 1989, he treasured it in his home in Yuen Long, Hong Kong. Now it is moved from Hong Kong to Shenzhen with him.

Kenneth Ko is willing to share with others. Twenty years ago, he occasionally auctioned an imperial robe with over one hundred years history in Hong Kong, intentionally invited a craftsman to inset it into two transparent glasses and placed it in his MY COFFEE for people to enjoy. Back to his home, he comes up with creative ideas. He uses a pulley gear as the moveable partition and background wall between dining room and gym, which is beneficial in two sides.

Walking in the house, what you see are historic antiques and rare art works. The ceiling decoration is from the Italian old building roof. Kenneth Ko thought the hand sketching on it was interesting and placing it in the ceiling could be a stroke of genius when buying it. The old wood screen which is transformed into TV wall describes the gentlemen manners of plum blossoms, orchid, bamboo and chrysanthemum. The wood cabinet and old clipping carpet from Turkey and the antique mirror from Venice are suitable and set off each other.

The out-of-print Chinese twelve flowers glass painting in the master bathroom and the four guardians sculptures in the guest room relate rich life experiences of the owner and Chinese tradition, which returns to true intention and is full of beautiful illusion of being in different regions and times every step.

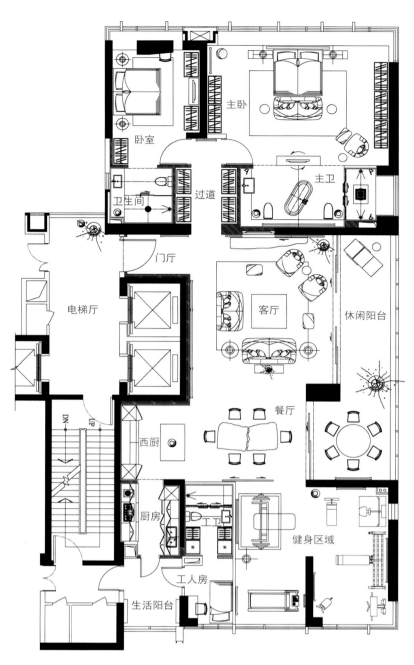

客厅油画《Temptation of ST Antony Par》，澳洲画家Noel Tunks作品，讲述了英女王伊丽莎白一世的故事，高文安对这幅作品异常喜爱，1989年高文安由澳洲买回后，一直珍藏在香港元朗的家中，如今随高文安的足迹从香港迁到深圳。

高文安一直乐于与人分享。二十几年前，偶然在香港拍得一件拥有过百年历史的龙袍，他即特意请匠人将龙袍镶在两块透明玻璃中，摆放于他的咖啡馆MY COFFEE供人观赏。移至此宅中，他再次发挥奇思妙想，辅以悬挂轮滑装置，用作餐厅与健身房之间的移动隔断，又可充当背景墙，一举两得。

信步居所中，所见都是历史悠久的老物件，或不可多得的艺术品：客厅天花的装饰是在意大利买回来的老建筑屋顶，购买时高文安只觉得上面的手绘很有意思，置于天花板的设计可谓神来一笔。改造成电视墙的老式木屏风，轻描梅兰竹菊的君子之仪；土耳其的木柜和拼接旧地毯，从威尼斯漂洋过海而来的古董镜；在这里都适得其所，相映成彰。

而主卧浴室中已绝版的中国十二花神玻璃画，和客卧的四大金刚雕塑坐镇，诉说着主人身上丰富的人生阅历和中国血脉，是回归本心，更充满了每走一步都穿越地域国家乃至时空的美好错觉。

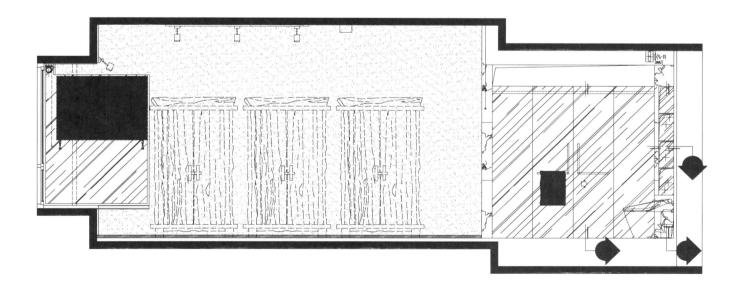

亚太名家荟萃·风靡大陆 | 047

亚太名家荟萃·风靡大陆 | 049

Master's feeling
大家情怀

The place of the heart is the feeling of the master. As is known, Kenneth Ko is nostalgic. In his home, there are many old things. The sofa leather of the living room is used for several decades. It is not because he doesn't have better. In fact he is not lack of valuable collections, but he likes using old things. In his words, there are emotions between them.

Kenneth Ko often says the best thing to invest is employing a world-class fitness instructor at the age of 50 to build up a sculptural body within three years. Today exercising is the biggest interest of his life. The portrait hanging in his home in Shenzhen is exactly his. It is his coach who inspires him to keep fit.

In particular, four of the eight rosewood back-rest chairs with mortise and tenon joint structures in the dining room are inherited from his father as family heirloom, and the other four are copied by craftsmen who are heavily employed by him in order to entertain guests, using the same rosewood. But they give him very different feelings. For nearly seventy years old Kenneth Ko, the four old chairs are his favorites, because they condense truly affections of the family.

The oil painting *Sima Qian Returning Home* hanged in the bedside of the bedroom is themed by Kenneth Ko and completed by the painter Yu Xiaofu for four years. The pursuit of classical Chinese culture is always his inner insistence. He thinks Chinese civilization is the essence which is worth to savor most. The reason why the Chinese design he advocates is different from other is that it has a profound cultural connotation.

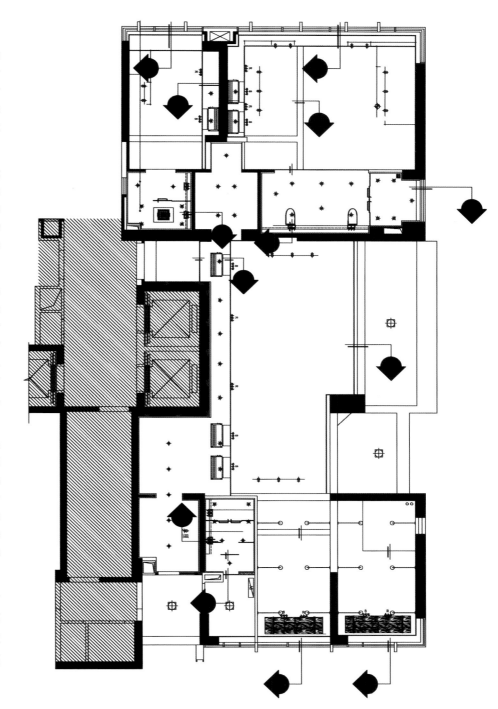

　　心之所至，大家情怀。都知道高文安念旧，在他的家里，可以找到很多旧物，客厅的皮沙发用了几十年不更换，并不是他没有更好的，事实上他并不缺价值不菲的藏品，但他仍然喜欢用旧的，用他的话来说，已经用出感情了。

　　高文安常说此生最值得的投资是50岁时聘请世界级健身教练，历时三年，锻炼出雕塑般的身形。如今，健身依然是高文安生活中的最大爱好，深圳家中的健身房悬挂的写真亦不是别人，正是他的教练，以此激励他对健身的坚持。

　　特别值得一提的是餐厅八把榫卯结构的红木靠背椅，其中有四把老椅子是高文安父亲留给他的传家之宝，另外四把则是高文安为了方便招待来客，重金聘请工匠仿制，用料是同样的红木，但给他的感觉不免依然天差地别。对年逾七十的高文安来说，那四把老木椅一直是他心头爱，因为其中浓缩了血浓于水的亲情。

　　次卧室床头悬挂的油画，是高文安命题，由画家俞晓夫历时四年倾心完成的《司马迁回乡》，对中国古典文化的向往，始终是高文安心念的执着。他认为华夏文明是最值得反复咀嚼的精粹，而他所倡导的中国设计之所以与众不同，是因为拥有深厚的文化底蕴。

亚太名家荟萃 · 风靡大陆 | 053

葛亚曦 / Kot

传承中的形与意
Inheritance of form and meaning
056

让生活向艺术再进一步
Make life one step closer to art
072

LSDCASA 创始人、艺术总监

葛亚曦作为LSDCASA创始人兼艺术总监，同时还身兼清华大学和同济大学软装设计班客座教授。他提出的"立于潮流之外，艺术构建生活"的理念，倡导生活方式的独特性，不盲从主流，忠于自我。

作品曾荣获伦敦SBID国际大奖住宅类金奖，入选Andrew Martin室内设计年鉴。2015年获SBID国际设计大奖（英国）住宅类金奖、亚洲+设计大奖赛（香港）最佳样板房设计大奖、现代装饰国际传媒奖年度软装设计大奖、第五届中国最佳设计最佳陈设艺术大奖、第十届金外滩大奖、第五届亚洲设计奖（台湾）银奖、美居奖中国最美样板间大奖、第10届金盘奖年度最佳样板房大奖等多项国际大奖。

传承中的形与意
INHERITANCE OF FORM AND MEANING

项目名称 ｜ 鲁能钓鱼台美高梅别墅
设计公司 ｜ LSDCASA
设计团队 ｜ LSDCASA设计一部
项目地点 ｜ 北京
项目面积 ｜ 660 m²

DESIGN CONCEPT ｜ 设计理念

In ancient Chinese poetry, "image" is a word which cannot be avoided. Endowing the "object" with "meaning" is constituting images into an organic, space-time distant and stratified pictures to make them cohere, contrast, foil or resonate to show the scene and to convey emotions.

　　在中国古代诗歌中，"意象"是个绕不开的词。对"象"赋予"意"，将一个个的意象按照美的规律，组成有机的、有时空距离的、有层次的画面，使其连贯、对比、烘托或共振，以展示场景、以传达感情。

In fact, it is the same as design. But the "object" of design "poem" becomes understandable symbol, element, article or rule, which is know as "form". The combination of forms conveys thoughts, philosophy and emotions.

This project is located in the middle axis of the third ring road with Temple of Heaven on the south. Its planning learns from the Forbidden City and the garden restores the Qianlong Garden. The form which LSD interprets is traditional Chinese style and the meaning is ritual sequence, respect and uniqueness. To convey the meaning by form or to reach the state by meaning, the first thing to do is inheriting at the same time communicating with the era.

实际上，设计也是如此。只不过，设计"诗"的"象"，成了可被解读的符号、元素、物件、规律，即我们所说的"形"，通过形的组合，去表达思考、哲学或情感。

鲁能项目坐镇三环中轴，天坛正南，项目规划效法紫禁城，园林灵感还原乾隆花园——LSD解读它的形，是传统，是中式，是"皇味十足"，LSD解读它的意，是礼序，是尊重，是不世俗。以形写意也好，以意达境也好，最首要的任务，是在做好传承的同时，对话时代。

"Erudite scholars come to talk with me, and among my guests there are no uneducated common people"—living room
"谈笑有鸿儒，往来无白丁"——会客

Entering the living room, *The Night Stream* by Mo Lijia will grasp your heart. The mountains under his pen is spectacular and extends endless images. The two dark and shallow couches make a contrast with the carpet. A set of pterocarpus santalinus official cap chairs is valuable, restrained and calm. The so-called "Great Music" is like that.

In the slant hall, traditional elements such as arhat bed, tureen and jade basin match with three durable plants of winter, pine, bamboo and plum blossom, which restores the entertaining and chatting scenes of literati in Wei and Jin Dynasties. The satin texture, primitive color of the pictures and colors of the curtains and carpets add beauty to each other.

The dining room continues the symmetric axis of the living room and shares the same luxurious scale. It is worth mentioning that the hand embroidery of the dining table chair back is weaved by the female embroider of Suzhou who uses gold and silk to spread inward gradually from the margin to make a dragon according to the patterns with layering covers, which stresses the crafts. The droplight is from Mathieu Lustrerie. The warm lights set off on the gilded tableware through the bronze and crystal surface. The shining lights create magnificent and valuable atmosphere.

Downstairs, the light and shadow of the six meters marble droplight pour down. The changeable texture on the wall is like exquisite landscape pictures. Following the grains of the landscape, the hazy lights are like thousand years cultural river and pour down onto a group of Taihu Lake stone.

The recreational area at the bottom of the stairs is exclusive for the man of the villa. The carpet is like a huge landscape painting which sets a tone for the whole space. Metal annular chandeliers and the metal round table contrast finely with each other. As the social space, there collects many precious memories from his travels.

步入客厅，莫里加的《夜溪》就将摄取你的心魄，他笔下的山峦，气象万千，延展出无穷无尽的意象。一深一浅两座长沙发，与地毯的跳脱形成对比。一组小叶紫檀南官帽椅，价值非凡却又内敛沉静，所谓"大音希声"也许正是如此。

一旁的偏厅，罗汉榻、三才碗、玉制盆器等浓重传统元素加入岁寒三友松、竹、梅的点缀，还原魏晋文人待客、清谈之场景，缎面的纹路、挂画的古朴色彩与窗帘地毯的用色相得益彰。

餐厅延仗客厅的中轴对称，享受同样的奢侈尺度。值得一提的是，餐椅椅背的手工刺绣，来自苏州绣娘一针一线织就而成——用金线、丝线两种线按纹样外缘逐步向内铺扎盘出龙图案，层层叠叠地铺就，十分重工。餐厅吊灯来自Mathieu Lustrerie，暖色的灯光透过青铜与水晶质面，映衬在描金漆餐具之上，粼粼微光，营造出大气、贵重的氛围。

顺着楼梯循级而下，6米长的云石吊灯光影如泄，石屏上幻化的肌理，犹如一幅幅精美山水画，顺着山水的纹路，透出朦朦胧胧的光如数千年的文化长河，倾泻于底下的一组太湖石之上。

位于楼梯底部的休闲区，专属于别墅的男主人，地毯仿若一幅巨大的山水画为整个空间的态度定调，金属环形吊灯，与金属圆几相映成趣。作为社交空间，这里更私藏了他游历四方遇见的珍贵记忆。

"Volumes of books can produce happiness"—study
"万卷皆生欢喜"——书房

In the *Delights and Pleasures* by Wu Congxian in Ming Dynasty, it talked about the ideal study that "If the study is deep enough, the threshold is flexural enough, the tree is sparse enough and the bed is spiritual, then volumes of books can produce happiness and there is no need to admire the fairy land". We can move all to reality.

The designer gets through the underground second and third floors to make a study. The 6.5 meters book wall can collect over 6000 books, which are the wisdom inherited by the scholarly family for generations.

Four treasures of study are set on the table, where the owner can put the casual poetry. What's more, the censer emits tranquil fragrance, which creates smoky atmosphere for the desk. With the accompany of bamboo and cypress, there are antiques and elegant arts on the shelf.

Four stones set in front of the desk are made of fossil woods. These precious legacies are formed by changed wooden parts after hundred-million years of trees were buried underground. They keep the texture and shape of trees with some polish. The different sizes are unique. They are placed on the primitive carpet as if born from it.

　　明代吴从先在《赏心乐事》中，谈到理想中的书房——"斋欲深，槛欲曲，树欲疏，榻上欲有烟云气，万卷皆生欢喜，阆苑仙洞不足羡"，我们将这一切搬到了现实中。

　　打通负二、负三层，以为书房，一面挑高6.5米的书墙，可藏书6000余本，是书香门第世代相传的智慧。

　　案上设有笔墨纸砚，安放主人随性而起的诗意，更以香炉焚香静气，为书案萦绕烟云。以竹柏作陪，架上藏纳古物雅玩。

　　案前放置四樽石墩，其材料为木化石，是上亿年的树木被迅速埋葬地下后，木质部分被改变而形成的遗世孤品。保留树木的纹理和形态，略加打磨，大小不一却更见情致。放置于古朴的地毯之上，仿佛从中生长出来一般。

"In the winter night when the guests come, you can treat them with tea instead of wine; the flames in the stove become hot, the water in the pot is boiling and the house becomes warm."—Zen tea

"寒夜客来茶代酒，竹炉汤沸火初红"——禅茶

Entertaining the guests by tea is the social etiquette of the ancient, can bring a quiet and meaningful artistic conception among friends and is regarded as an elegant thing. Therefore there is a tea room in the first floor underground, which offers a better place for a chat and a cold night.

It restores the tea room of Su-Shi when he stayed in Shushan Mount, facing the bamboo in the yard, "there is a hut with some bamboo and stones, where you can make tea and treat the guests". There are only two seats in the tea room, where you can enjoy the bamboo at the distance and make poem closely, which is "a cup of tea can acknowledge your friends".

以茶待客，乃古代人情交际的礼节，它为友人之间带来一种清幽隽永的意境，更被视为风雅之事。故此，负一层设禅茶室，供长日清谈、寒宵兀坐。

这里还原了苏轼闲居蜀山时的茶室，正对院竹，"茅屋一间，修竹数竿，小石一块，可以烹茶，可以留客也。"茶室仅设二席，远可观竹，近可对诗，是为"一盏清茗酬知音"。

"You can play plain piano and read buddhist sutra"—Piano room

"可以调素琴，阅金经"——香道琴房

Plying the piano is a kind of elegant habit of literati. The piano room in the first floor underground is a place to play and meditate. The ancients regard piano as character with ten kindnesses, ten commandments and five no-playing things such as not playing in the mortal life, not playing to common people, not playing without clothes, and have high demands on environment and themselves, such as clean floor, perfect state, elegant house and tasty fragrance. Therefore, there are elegant seats, censer in the piano room with poems in the pictures. Through the yarn curtains, the plantains intersperse in the yard, which fits the leisure artistic condition of "under the plantains in the rain, you can play the piano".

斋中抚琴，也是文人的一种雅好。负一层的香道琴房，是抚琴、冥思的安静所在。古人视琴如格，有十善、十诫、五不弹，如于尘市不弹、对俗子不弹、不衣冠不弹等，对环境及自身的要求都极高，或地清境绝，或雅室焚香。故此，香道琴房设雅席、设香炉，以诗词入画，透着纱质窗帘，院中芭蕉隐隐点缀，正应了"芭蕉叶下雨弹琴"的闲适意境。

"Harmonious with diversity, each is in the proper place"—living room
"和而不同，各得其所"——居室

Large area of gilded paintings and quiet and tranquil dark color are the main tone of the master bedroom with coppers interspersing inside. Through the hollow wood screen, the sunshine falls on the heavy green branches and leaves, which adds yard fun to the bedromom.

"Opening the window, you can see the moon and the wind blows over the bed", since the Han Dynasty, a couch is necessary for refined scholars who use bamboo couch, stone couch and wood couch to represent their loftiness and quality. The arhat couch besides the master bedroom can place the leisure body and mind, where you can display historic works, read books, play the piano and sleep with the fragrance. The antiques and elegant works on the desk add some elegant fun.

What is called gaining in autumn and collecting in winter, the guest room matches sedate and quiet dark gray with indifferent and plain dark green. The entire space is based on this tone. Metal furnishing articles echo with the bronze droplights, which presents a depositing beauty of years in the primitive and indifferent atmosphere.

The third floor is children's secret amusement park, with the boy's room at left and girl's room at right dividing by Yin and Yang.

Boy's room pursues fashion with the main tone black and white interspersing in large display area and details in the space. The carpet with casual and abstract lines create a lively and vivid atmosphere in the space.

Girl's room matches elegant gray with quiet purple. The metal texture of table edges and droplight echoes with each other. The curtain combined by red green and white yarn dances with the wind, whose dream can it contain?

大面积的金箔画与静气内敛的深色系为主卧渲染基调，铜器的光泽穿插点缀，透过镂空的木质屏风，漏下了点点滴滴的夕阳余晖，映照在浓绿的枝叶上，将卧室书写出了庭院的意趣。

"推半窗明月，卧一榻清风"，自汉末以来，文人雅士必备一榻，以竹榻、石榻、木榻来表示自己的清高和定性，主卧一侧的罗汉榻，用以安放闲适的身心，展经史、阅书画，或倚坐抚琴，或睡卧闻香，案上搁放的珐琅古物与雅玩，更增添几分风雅妙趣。

所谓秋敛冬藏，客房用深灰的沉稳静默，搭配墨绿的淡泊质朴，整个居室的秉性就在这样的基调中游走。金属质感的摆件，呼应青铜吊灯，在古朴的淡泊中透着一丝岁月沉淀的美感。

三楼是孩子们秘密的游乐园，以阴阳分布的男左女右，安置了男孩房、女孩房。

男孩房追求时尚的跳脱，以黑与白作为主色调，穿插在空间的大展示面与细节中。线条随意抽象交织的地毯，铺就整个空间活泼生动的气氛。

女孩房以优雅的灰色搭配恬静的紫色，金属的质感在边几与吊灯之间遥相呼应。豆沙绿与白纱组合的窗帘，在微风下细细撩动，能否装下谁的一帘幽梦？

让生活向艺术再进一步
MAKE LIFE ONE STEP CLOSER TO ART

项目名称 ｜ 上海万科翡翠滨江300
设计公司 ｜ LSDCASA
设计团队 ｜ LSDCASA事业一部
项目地点 ｜ 上海
项目面积 ｜ 300 m²
摄 影 师 ｜ 啊光

DESIGN CONCEPT ｜ 设计理念

The inheritance and possession of handmade luxurious furniture happens to a few people. Industrial revolution brought mechanized and monotonous implements which are used in daily life. So solemn hand-polishes and unique delicate textures are more valuable.

　　手工奢华家具的传承与拥有，历来仅发生在少数人中，如今尤甚。工业革命带来机械化、单调的器物，并充斥于生活常态。也正因如此，手工打磨的庄重和独一无二的细致纹理，才越显珍贵。

Vanke Emerald Riverside is located in the core area of Lujiazui riverside with 1.4 kilometers luxurious skyline. The valuable position must be treated seriously. So LSDCASA traces back to aesthetics of Raphael era, stresses thoroughly tempered handworks and injects art into it, which endows Emerald Riverside 300 house type with an inherent value.

Metal cabinet piled by metal lines gives a different visual impact at first sight. The big abstract decorative painting as the background creates an artistic and pioneering tone.

Designs of the living room stress comfortable functions and every detail which can presents the tone. Light L-shaped sofa, selected Andrew Martin fabrics, mazy droplight bans, combined crack tea table, stone cut desk, chairs of 1970s and sculptural side table try to embody artistic tones within orders.

When lights on, the magnificent landscape of Lujiazui and blurred scenery of Huangpu River are brought into interior by the 270° French window, and shine beauty with the artistic lights. The valuable position is extraordinary and outstanding.

Dining table, chairs and sculpture platform refuse traditional geometric sharp angles. A lot of curves create manual aesthetics and echo with the top droplight modeling. Bamboo joints of the dining chair feet are concise and natural with ten million times hand polishes backward, which is precious.

万科翡翠滨江，坐镇陆家嘴滨江核心区，占尽1.4公里繁华天际线，如此贵重地位，必须庄重对待。于是，LSDCASA追溯前拉斐尔时代的审美，重用手工的千锤百炼，艺术贯穿其中，让翡翠滨江300户型的华贵价值，与生俱来。

金属线条堆叠造型的金属柜，给进入空间的第一眼带来不一样的视觉冲击，以大幅抽象装饰画为基底，交叠出艺术、先锋的格调。

客厅设计更加偏向舒适的功能性，但仍不放过任何一个释放格调的细节。浅色L型沙发，精选Andrew Martin面料、迷宫布阵般的吊顶灯带、组合式裂纹茶几、石块般切割的书桌、20世纪70年代的书椅、雕塑形式的边几，一切都试图在规矩中体现艺术性。

待到华灯初上，陆家嘴的时代盛景和黄浦江的迷离景致，被270度环伺的落地窗完美吸纳，与空间中如艺术般交织的灯光相辉映，卓然地位，自此非凡。

餐桌、餐椅与雕塑台拒绝传统的几何利角。大量的曲线打造手工美感，与顶面吊灯造型相互呼应。餐椅脚的竹节式样，简约自然，背后需要千万次的手工打磨，这份用心弥足珍贵。

Compared with chic dining room, master bedroom tries to resonate between comfort and art. The tone of purple, gold and green manifests luxury and nobility. The individual chest of drawers and tall cabinet are less decorated, which is appropriate. The carpet is like a decorative painting, which largely improves the decorative sense of the space.

Different bedrooms are endowed with different themes. What remains unchanged is the inherited artistic sense and honest value. Light gray and dark green are the main tone, which is sedate and unobtrusive. Matte surfaces and historic textures are not noisy and form their own styles. If the main tone is black, white and gray, then the whole space is restrained and solemn. Wallpapers with abstract patterns match with art installations, which creates extraordinary artistic sense within restraint.

相较于别致的餐厅，主卧力图在舒适与艺术之间达到共振。紫金绿的色调显奢显贵，个性的五斗橱与高柜稍作点缀，不多不少。脚下的地毯仿佛是一副装饰画，大面积地提升整个空间的装饰感。

不同的居室，赋予了不同的主题，不变的是一脉相承的艺术感和以诚相待的贵重。浅灰与深绿形成主色调，在视觉上稳定而不突兀。用哑饰面和有历史感的纹理，不喧哗，自成格调。主色调为黑白灰，整体空间内敛而庄重。抽象图案的壁纸与艺术装置相映衬，在克制中跳脱出不凡的艺术嗅觉。

凌子达 / Zida Ling

KLID达观国际设计事务所 设计总监、创始人

凌子达，1973年出生于台湾高雄，1999年毕业于台湾逢甲大学建筑系。2001年在上海成立"达观国际建筑室内设计事务所"，致力于建筑室内空间设计领域。2006年出版个人作品集《达观视界》。2009年取得法国Conservatoire National des Artset Metiers建筑管理硕士学位。

坚持"达者为新，观之有道"的设计宗旨，用豁达的胸襟看世界，以开阔的视野带来新的创意和想法。2014年荣获德国红点设计大奖之红点奖，以及荣获德意志联邦共和国国家设计奖，还有英国BLUEPRINT设计大奖等。2015年德国设计大奖，德国红点奖特别提名，美国ARCHITIZER A+ AWARDS最佳荣誉奖以及美国IDEA室内设计大赛优秀奖，同时获得新加坡GOOD DESIGN 设计大奖Gmark，英国aisa proerty awards 最高荣誉奖，获得设计界奥斯卡奖之称的英国伦敦安德马丁(ANDREW MARTIN)国际室内设计大奖等。登顶意大利A'DEDIGN AWARD组织的2016-2017年度世界领先设计师（室内空间与展览设计类）排行榜第一名，2016年度世界设计排行榜中国第一名，荣登2016-2017年度世界最佳设计师排行榜全球第四名等。

豪宅匠心 独具风韵
MANSION WITH UNIQUE CHARM

项目名称 ｜ 之江九里
设计公司 ｜ KLID达观国际设计事务所
设 计 师 ｜ 凌子达、杨家瑀
项目地点 ｜ 浙江杭州
项目面积 ｜ 850 m²

DESIGN CONCEPT ｜ 设计理念

This is a six-floor single-family villa with elevator. The overall layout of the building pursues leisure and art functions. The most special place of the villa is there are two floors underground and four floors on the ground. Usually the basements leave gloomy, damp impressions on people without enough lights and air circulation. To break the general impressions and ideas of inherent basement, we boldly reorganize and design the space and partition and transform the whole. We keep the main structure, remove all walls and floors and remain the original structure of the architecture and reconstruct the floor in higher space to make spaces echo each other. Through this way, we specially add bridge and corridor in tall parts to shorten the distance between people so that they can connect and communicate.

这是一栋6层楼的电梯型独栋别墅，对于建筑整体的布局，主要是以休闲与艺术的追求功能性为主。这栋别墅最特别的地方在于地下室有两层，地上有四层，一般地下室给人印象是比较阴暗、潮湿、光线不足，空气不流通。为了打破一般人对地下室固有的印象与想法，我们大胆将空间及隔间做了重新规划与设计并且整体做改造，让主体结构不变，把墙体、楼板全部打掉，保留了建筑原有的结构柱，在挑空的部分楼板重新开口，让空间能上下相互呼应。透过这样的方式，我们特别在挑空的部分加入了桥、回廊等，让人跟人之间没有距离，能够互相串联与沟通。

Most especially, we boldly design a guest room in Basement one. Large areas of glasses bring natural lights into interior, which makes the basement brighter. We focus on three aspects to endow the basement with life and vigor. First, in order to give users different soul feelings, we use large areas of window to make them feel the air flowing and comfortable, which defines transparent senses. Second, we use tall height to define the space and bring lights into interior with wide view to make the residents feel the holiday environment. Last, we adopt landscape sense, use the layout of plants and create a different space sense through layering and intersection.

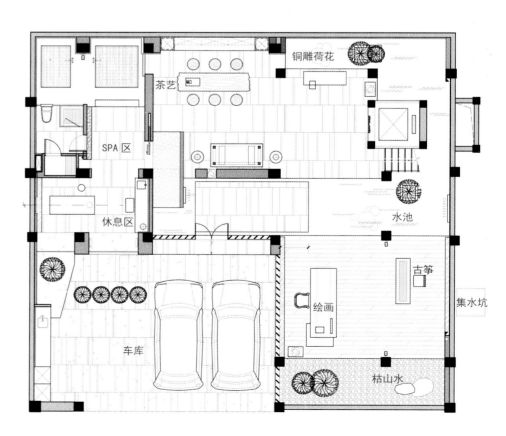

　　最特别的是在B1层，我们大胆设计了一间客房，透过大面积的玻璃来增加自然光源能够进入到室内，让地下室感觉更加明亮。将地下室使用空间赋予具有生命与活力来作为我们这次设计的三个方向。首先，为了让使用者在这个空间使用上能有不一样的心灵感受，我们运用大面积的开窗，让室内空间在使用上能够感受到气的流通与舒畅，极具通透感的界定。其次，运用挑空的高度来界定空间，光线导引进入室内时有宽广的视野，让使用者能有像度假般的环境。然后，植入枯山水的干景，运用植物景观的布置，透过错层与互相交错达成一个不一样的空间感。

The first floor is the living room and dining room. The second floor has three suits of rooms and a second master bedroom. The third floor is master bedroom. And the forth floor is a loft. Basement two has SPA, tea room, gym and recreational area with pool and garden inside. This kind of method endows the entire floor with Oriental cultural elements such as four arts.

一楼作为客厅与餐厅，二楼设计有三间套房、一间次主卧，三楼则是主卧房，四楼是阁楼。特别在B2层设置SPA区、茶居室、健身房、休闲运动区，室内增添了水池、园林等，透过这样的手法，让整层空间带有东方琴棋书画等文化的元素。

戴勇 / Eric Tai

戴勇室内设计事务所 董事长、设计总监　卡萨艺术品公司 董事长

　　戴勇，生于1971年，24岁开始从事室内设计，1997年处女作"深圳天威数据写字楼"荣获中国室内设计学会佳作奖。2004年创立个人设计公司，同年作品"佳兆业桂芳园样板房"荣获海峡两岸四地设计大奖一等奖。2005年获颁"深圳十大设计师"称号，获奖作品"深圳华尔登府邸"发表于台湾《室内》杂志。2009年获颁"中国十大设计师"称号，并于当年获选英国Andrew Martin室内设计奥斯卡全球优秀室内设计师，两件作品入选Andrew Martin获奖作品集，成为深圳首位获此殊荣的设计师。2012年受聘清华美术学院高级室内设计研修班授课教师，担任室内设计及陈设设计的实践导师。

典雅空间 奢华风尚
ELEGANT SPACE, LUXURIOUS STYLE

项目名称丨惠州君御复式样板房
室内设计丨戴勇室内设计事务所
艺术陈设丨戴勇室内设计事务所
项目地点丨广东惠州
项目面积丨241 m²
摄 影 师丨江国增
主要材料丨意大利孔雀蓝玉石、白金沙云石、直纹柚木饰面索色、玫瑰金镜面不锈钢、米白色皮革、夹丝镜等

DESIGN CONCEPT | 设计理念

This project uses precious Italian peacock blue jade as TV background wall, which satisfies functional needs and is a technique of expression which achieves sensual attractions from the perspectives of design. In the luxurious and modern space, the applications of materials are very rich. The space modeling pursues tight and clear layering. Visual differences from ceiling, partition to design of a single wall give people resplendent and magnificent feelings, which reflects the needs of wealthy class. It is interpreted as a diversified style. From modeling, materials and soft decorations, people call it Chinese Art Deco. However, the choices of colors, materials and painting patterns manifest sedate temperament of the owner.

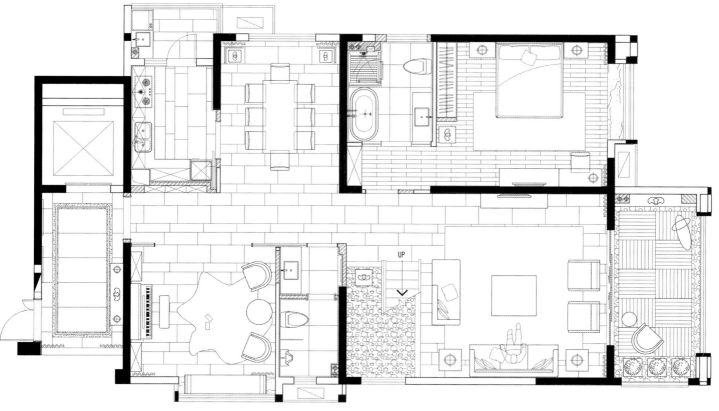

本案运用珍贵的意大利孔雀蓝玉石做电视背景墙，既是功能的需要，从设计角度分析也是为实现感官吸引力的一种表现手法。在这个奢华的现代空间中，材料的运用非常丰富，在空间造型上也要求层次紧密分明，无论从天花到隔墙甚至到单面墙的设计上显示的视觉落差，给予人的感受性都是金碧辉煌富丽堂皇，体现财富阶层的一种需求。它甚至被解释为多元化的风格，单从造型材料及软饰的繁芜上有人甚至将它称为中式Art Deco，然而从色彩选择、木料偏好及挂画图案的选择上依然体现出主人的稳重气质。

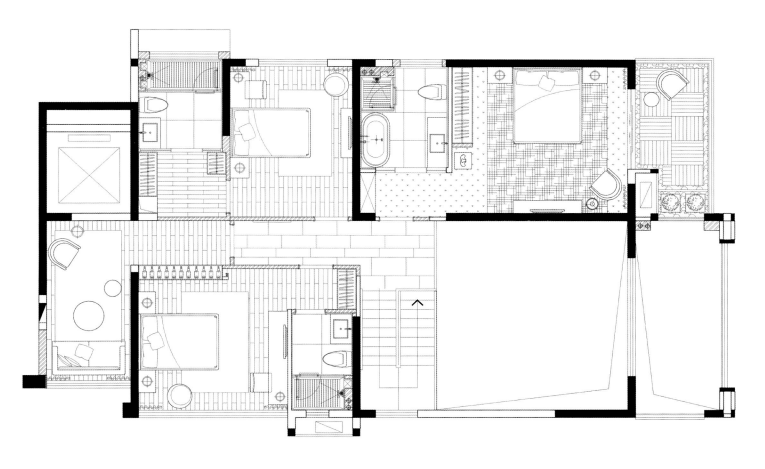

刘卫军 / Danfu Lau

PINKI（品伊国际创意）品牌创始人、董事长兼首席创意总监

刘卫军，美国（国际认证及注册协会）注册高级设计师、中国首批注册国家高级室内建筑师、中国建筑学会室内设计分会全国理事及深专委常务副会长，IFDA国际室内装饰协会理事、广东省家具商会OIDE国际品牌与设计交流中心设计委名誉会长、ADC设计研修院导师、ADC设计师资质认证委员会主任评委、清华大学美术学院陈设艺术高级研修实践导师及全国高级陈设艺术设计导师。

首登《亚洲新闻人物》的中国设计师、2002年中国人民大会堂推行发布陈设艺术配饰专业发展第一人、2002年博鳌家居论坛中国室内设计师风云人物、CIID学会第一个代表中国设计界赴韩国首尔担任"第四届亚洲室内设计联合会年会暨国际室内设计学术交流会"演讲人、CIID学会第一个亚洲室内设计论文奖获得者、CIID学会第一批著书立作的设计师、中国室内设计20年推动大奖、SIID深圳市室内建筑设计行业协会首届副会长及2009年中国时代新闻人物。

大艺术家·厢
THE ARTIST AND WING

项目名称｜广州金茂府5#D-2示范单位
设计公司｜品伊国际创意产业集团有限公司
设 计 师｜刘卫军
项目地点｜广东广州
项目面积｜200 m²
摄 影 师｜曾朗、黄明德
主要材料｜大理石、木饰面、木地板、墙布、金属、艺术玻璃、皮革等

DESIGN CONCEPT ｜ 设计理念

A flower and a paradise, a grass and a world, a mind and a peace, a wing and a pure land

一花一天堂，一草一世界，一念一清净，一厢一净土

Wing and calmness
【厢】容

If you come, I will be here waiting for you, making tea for you and listening to your favorite songs in a land of calamus. Taking up yesterday's book, I'm absorbed in the warm and quiet atmosphere. The red in the distance outside the window adds flexibility into the quiet space and warms people.

你来，我就在这里等你，为你沏茶，菖蒲做兴，聆听你最爱的歌曲，俯身捧起昨夜的书，沉浸在温婉而静谧的空气中。远处窗前的那一抹红，让安静的空间多了份灵动，还能给人以暖意。

I don't need a very large space. When free, I can sit quietly by the window to read a book, be in a daze quietly, have a cup of tea or enjoy the red which manifests the vigorous and wonderful life, which makes me full of joy and confidence to life.

我不需要很大的空间，闲来无事的时候，可以一个人静静的坐在窗前，读一本书、或静静的发呆、或品一杯茶，欣赏眼前这昭示着生命的旺盛与精彩的一抹红色，让我对生活充满了喜悦与信心。

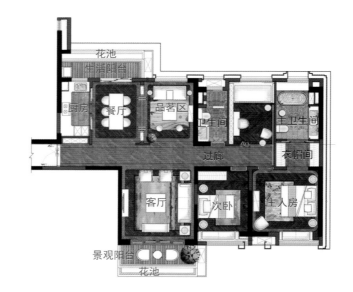

Wing and mat
【厢】席

The beauty of calamus is static with dynamic and full of elegant flavors. As Su-Shi put: "Although it is not flourish, the nodes and leaves are solid and thin. The roots intersect. Placing in the table, it is healthy and enjoyable." Since Tang and Song Dynasties, calamus came out of the steam side into the desks of literati. We also use the beauty in our life. Although it is not flourish, the nodes and leaves are solid and thin with an elegant posture, which can bring wild and leisure fun into everyday life. The noises of the world seem to be isolated outside by the red.

菖蒲野草之美，静中有动，文雅气息。东坡文中所言："虽不甚茂，而节叶坚瘦，根须连络，苍然于几案间，久而益可喜也。"从唐宋开始，石菖蒲这种植物，就走出溪头涧畔，成为文人案头清供。我们也将这种野草之美，植于我们的生活，虽不甚茂，而节叶坚瘦，姿态秀雅，为日常带来野逸之趣。尘世的喧嚣嘈杂，仿佛被这个一厢红色隔绝到了身外。

Wing and wine cup
【厢】觞

In the gray table there is a white, which is delicate, pretty and elegant in the red painting with lotuses. Inviting friends to taste delicious dishes cooked by the wife, though there is no fish and meat, elegant food and environment can be their favorite. Without grand music, drinking wine and writing poem can be enjoyable.

灰色的桌面上中竟有这样一抹白，一株株荷花在红色画中，更显得清秀、洁雅、邀友人至家中，尝尝爱人烹饪的可口菜肴，虽没有鱼肉的畅快，但吃的清雅与环境也是友人的挚爱。虽无丝竹管弦之盛，一觞一咏，亦足以畅叙幽情。

Wing and reading
【厢】阅

The wise loves water and the benevolent loves mountains. The wise is dynamic and the benevolent is static and as steady as the mountain. The study is the host's favorite place where he can enjoy light sanders around alone. The landscape painting behind makes the study quiet under the foil of lights.

　　智者乐水，仁者乐山；智者动，仁者静，仁者像大山般坚守不动，书房是男主人最爱的地方，独享淡淡的檀木香充斥在身旁，背后的山水画，在灯光的衬托下，书房显得这样的安静。

Wing and rest
【厢】息

A cup of tea spends ten minutes and a short span is ten instants. Time carries tranquility. The eternity you want is just the heart stilled at this moment, which is tranquil. It's just life. The thoughts convey the purest simple beauty inside. There is a wing which carries thoughts, elegance and love.

　　一盏茶十分钟，一弹指是十刹那，时间承载着宁和，你想要的永恒也不过是这一瞬静止的心，波澜不惊。生活不过如此，一厢的思绪寄托着内心最纯净的简单美好，有个可以寄托思念与雅致，承载爱的地方。

孟也 / Ye Meng

孟也空间创意设计事务所 设计总监 渡道国际空间设计(北京) 创始人

孟也，著名别墅设计师，是装饰设计行业中炙手可热的新锐独立设计师。他一直坚持"创新是一种信仰"的设计理念，认为设计需要创新，才能永葆设计的生命力和调性；信仰也尤为重要，坚持真诚待人和把项目做好的信念。入行多年，他最重要的设计心得是抛开形式做实质，从每一个客户的切身利益考虑，深入客户需求内核，设身处地为别人的幸福做出努力，"设计最终是为人服务的"。

近两年取得的卓越成绩有：2014年美国莱斯杂志中国室内设计年度封面人物、2014年金堂奖别墅十佳设计师、2014北京国际设计周"十二间"公益展特邀设计师等。

和谐惬意家
A HARMONIOUS AND COZY HOME

项目名称 ｜ 远洋别墅
设计公司 ｜ 孟也空间创意设计事务所
设 计 师 ｜ 孟也
项目地点 ｜ 北京
项目面积 ｜ 1500 m²
主要材料 ｜ 布艺、木材、大理石、壁纸等

DESIGN CONCEPT ｜ 设计理念

Gorgeous Western space art trend seems to drop the curtain, and Chinese new residential space design merges quietly. With a more indifferent and rational mind, it builds and experiences the high quality life of modern people's views.

Living here, without being bound by retro marbles of the façade and the European arched window, you can do as you wish, can be wayward not like the European architecture with a same outside and inside and don't need to please the society. Because this is the home, the only place where you can do whatever you want in the world. "You can do anything good with your loved ones at home."

华丽丽的西方空间艺术潮似乎落下帷幕，中国新的居住空间设计已悄然兴起，在更加淡然、理性的心态下，构建、体验现代人视野下的高品质生活诉求。

居住在里面，不被外墙复古风格大理石和欧式拱窗所束缚、随心所欲，任性地不用和建筑的欧式表里如一，不再需要讨好社会，这里是家，是这世界上唯一可以为所欲为的地方。"可以在家与爱的人做任何美好的事物"。

Floating in the air, the space is extremely free. Lights from the lined up southern windows of the living room, dining room, sitting room and tea room and the transformed sky-light windows shine in the uncomplicated decorations, which is just right and appropriate.

Space design is not only the carrier of furnishings, furniture and soft decorations, but also the emotional sustenance. Convenient life, deliberate act, casual delight and deposit mind are here, which is the eventual and simplest life goal of the resident.

飘然在室内，空间是极自由的，光线从客厅、餐厅、起居厅、茶坊一字排开的南向窗户和后改造的顶窗照射进来，照射在并不复杂的装修上，恰到好处。

空间设计，不止是陈设、家具、软装的载体，也是情感的寄托。生活的便捷、行动的从容、欢畅的随性、心绪的沉淀都在这里发生，也是居住者最终、最简单生活的目的。

张成喆 / Alessio

IADC涞澳设计 创始人、首席创意总监

  毕业于鲁迅美术学院环境艺术系，是目前中国炙手可热的室内设计师。其设计以冷静、简洁并富有创意的设计而著称，他认为室内设计如同服装设计，材质、色彩、造型可以随流行趋势而变，但逃脱不开其最核心的价值与理念。他的设计作品常发表设计类媒体如《安邸AD》，《家居廊》等。2006年出版个人作品集《喆思空间》。

  设计之外，他更是一个生活艺术家，喜欢随性洒脱、飘逸不羁的生活。他认为，要做一个好设计师，首先一定要做一个好的生活家，爱生活，才能有更多好的作品，这就是他的设计秘诀，也是他的生活哲学。

人和房子，需要慢慢培养感情
HUMAN AND HOUSE NEED TO CULTIVATE RELATIONSHIPS SLOWLY

设计公司｜IADC涞澳设计
设 计 师｜张成喆
项目地点｜上海
项目面积｜500 m²
摄 影 师｜金霑，SELLN LARS JOHAN FREDRIK
主要材料｜实木地板、大理石、铜、乳胶漆等

DESIGN CONCEPT ｜ 设计理念

The designer Alessio is definitely a perfectionist. It takes him more than five years from buying this villa to design to decorate to live in. He still has fresh memories about the repeatedly changed design drawing and construction details. Perhaps long time "being together" is exactly the way to cultivate relationships with this house.

设计师张成喆绝对是个完美主义的崇拜者，从买下这套别墅到设计、装修、入住，足足花了他超过5年的光阴，一改再改的设计图纸和施工细节他还记忆犹新，而长时间的"相处"也许正是他和这栋房子培养感情的方法吧！

Living in Europe for half the year, Alessio thinks he is a "very homesick" person who even takes a full set of bedding. Every time visiting Paris, he would live in the same area and go downstairs to the market to buy fresh food to fill up his refrigerator. He thinks that sense of security and freedom are exactly key words about home.

So for homesick him, moving from the Xinhua Road apartment where he had lived for nearly ten years to this newly decorated villa is difficult. Several years ago, he and his wife bought this villa located in Pudong Vanke Emerald Riverside. At first he had no feelings about this big house and never thought of living here. After being bought, the house has been set aside for a while. Later he decides to transform it into his second residence in Shanghai.

Then he begins designing and space planning and never thinks of lasting for five years. The construction organization was changed twice and the design drawing was changed countless. Alessio says that it is clear to give clients their schemes at the beginning. But when it comes to his own residence, without limitation about time and budgets, he can research constantly. After a longer time, his inside and lifestyle are changed.

一年中有半年时间旅居欧洲，设计师觉得自己是个"特别恋家"的人，也许会带着全套床品旅行，每次到访巴黎只住同一区，一到公寓就马上下楼找菜场，用新鲜食物塞满冰箱，他认为，安全感和自由，确实是家的关键词。

因此，对恋家的他来说，要将感情从生活了近10年的新华路公寓，慢慢转移到这套新装修的别墅，是一件不容易的事情。几年前，他和太太Eva买下了位于浦东万科翡翠滨江的这套别墅，最初看到这栋大房子时他完全没有感觉，没想过要来居住，买下之后，就这么搁置了一段时间，才决定改造作为上海的第二住所。

于是他开始做设计、布平面，未料经历了快5年时间，施工单位都换了两家，设计图纸更是更改了无数份，他笑言，"给客户的方案，一开始就会想得很清楚，但自己居住的地方，没有时间和预算限制，就能不断钻研；经过的时间越长，内心和生活方式都在变化。"

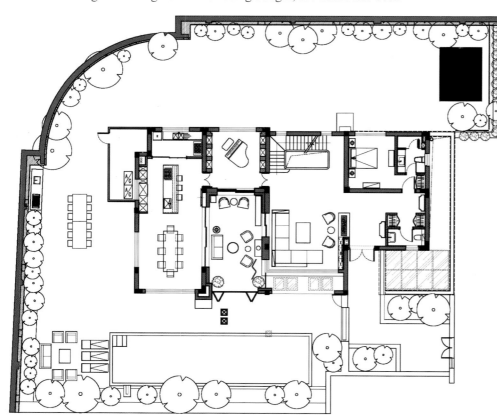

He always pursues spacious and flowing sense of space. So he skillfully transforms the typical villa with many rooms which is suitable for a large family to live in into a suburban living space for a dink couple who pursue freedom. He removes the partition which makes the second floor closed and rigid and creates a new layout which has two bedrooms at each side of the corridor and a bright and spacious open study in the middle. The first floor has a swimming pool specially made for Eva who loves swimming in the yard. The sunny yard links the living room with the hall and becomes guests' favorite side hall after roofed. Few people use the entrance of the original living room. Everyone loves to walk along the wood floor burnt hot by sun passing through the green swimming pool, enter into the warm and comfortable side hall from the tall and spacious French window, and go to the kitchen to have a glimpse of the host who is preparing French cuisine.

Every seemingly random and natural detail indoor is a result of thoughtful thought of the owner. In order to balance large scale gray wood doors, European wood lines are added into the doors, which is not complicated. The thickness, length and radian of every line are considered for a long time. Brass locks of every door are brought by the owner from second-hand market in Europe and are in the same color and different patterns. In the hall, sofas respectively from Paris flea market are matched into a set naturally and are specially decorated with coarse linen to neutralize the exquisite European style. The walls of the side hall are coved with dry old pine boards. Even the skylight leaks water, he is not worried about clapboard transformation. Alessio expects to see the traces of time after a long time when the pine boards fade or gradually have paint dropped under the sunshine. This is exactly the natural and relaxing feeling.

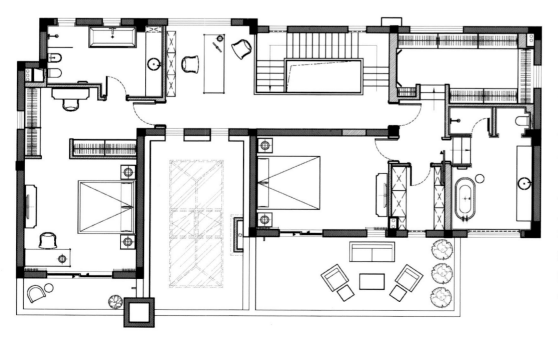

素来追求宽敞流动的空间感，设计师妙笔一挥，将原本有很多房间、适合一大家人居住的典型别墅，改造成这对追求自由的丁克夫妇的郊外生活空间。移去使二层空间闭塞死板的隔墙，形成了现在走廊两头，各有一套卧室，中间是明亮宽敞的开放式书房的格局。一层的院子里有特别为喜爱游泳的Eva打造的泳池，连接客厅和餐厅的阳光庭院，加上玻璃屋顶后成为客人们喜欢的侧厅，原先的客厅正门反而少有人用，大家都喜欢踏着被阳光晒得温热的木地板走过一池碧波的泳池，从高大敞亮的落地玻璃门进入温暖舒适的侧厅，还可以去厨房一窥热爱烹饪的男主人正在准备的法式美食。

室内每一处看似随意自然的细节，都是主人反复推敲的结果：为了平衡大尺寸的灰色木门，门上加入了欧洲风格的木线条，却不过于烦琐，每一根线条的粗细、长短、弧度都推敲许久；每扇门上的黄铜锁件，都是主人从欧洲各地的二手市场上淘来的老件，颜色一致而式样各异。厅里，在巴黎跳蚤市场分别淘来的沙发，却几乎天衣无缝地配成了一套，还特意选配了粗糙麻布，中和过于精致的欧式风格。侧厅墙上是干燥处理的老松木板，即使天窗漏水也不担心护墙板变形，他还尤其期待着日子久了之后，在阳光的照射下松木板开始褪色，或者逐渐有油漆剥落，显示出时间的痕迹，这才是自然的令人放松的感觉。

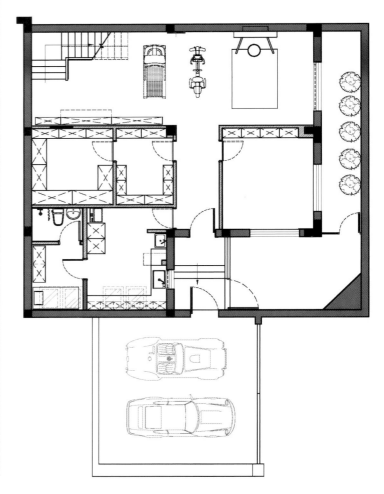

For home, Alessio insists on the principle of "getting along slowly". As for how to get along, the owner smiles and says everything should start from the kitchen. He tries his best to carry back a set of European copper cooking stove from Paris. There is no modern electronic originals, and the baking time and temperature are relied on experience. Every cook is an interesting adventure. The side wall hangs a set of exquisite kitchen ware. Mellow wood handle with copper matches with handmade brown leather sheath, which is a feast for the eyes. Living here, he cooks everyday for himself. He fries fresh mussels with butter, parsley, onions and thyme. With a cup of iced white wine with fruit fragrance, he can leisurely enjoy a healthy and delicious meal at the dining table near the pool. When the inspirations come, he can move the easel to the sunny living room to sprinkle his minds. "In fact I don't need very large spaces, but I especially need a space in my mind and heart which belongs to myself. Whether it is ups and downs or tranquility and sadness inside, it is totally your own world."

Alessio says frankly: "Now being here, I don't completely regard it as a home, feeling like a travel which is not much familiar and somehow a little strange. But there are a lot of fun to be explored. This is the house. How long you live in, how long you adjust yourself." He often brings some old collections. When he feels wrong about the place he puts them, he moves them and then brings other things to try. "My own home must have my favorite style. The home in Xinhua Road is my favorite style ten years ago, which is avant-garde, fashionable and concise and stresses black and white contrast. With time going by, now I prefer neutral and diversified style. It looks like that we spend five years creating this home, in fact many things here are accumulated slowly by more than ten years time."

Alessio also has a favorite thing, that is a set of original cartoons *The Adventures of Tintin* issued in 1930s. He spent many years in French second-hand market searching them one by one and collects them around himself. Sometimes bringing them to have a look, smelling the familiar smell can make him sleep sound. "The moment I bring them here is the time I regard here as my home." He smiles and says.

对于家，设计师坚持的原则是"慢慢相处"，至于如何相处？主人笑称一切要从厨房开始。他想尽办法从巴黎运回来一套纯铜的欧式炉灶，烤箱里没有任何现代的电子原件，全靠经验决定烘焙的时间和火候，每一次下厨都是有趣的冒险。侧面墙上挂着一整套精美的各式厨具，圆润的镶铜木柄配上手工缝制的棕色皮套，光看着已经赏心悦目。住在这里的时间，他每天都会自己下厨：新鲜贻贝，用黄油、欧芹、洋葱和百里香一炒，配上一杯果香怡人的冰镇白葡萄酒，就可以在泳池边的餐桌旁悠悠享受健康美味的一餐。或是在灵感来临的时刻，把画架搬到洒满阳光的客厅，在纸上挥洒才情。"其实我并不需要很大的空间，但是特别需要脑子里、心里的空间，有那么一块地方，是你自己的，里边跌宕起伏也好，安静伤感也好，完全是你自己的世界。"

设计师坦言："现在到这里来，其实还没有完全当成家，有点旅行的感觉，没有那么熟悉，有一点点陌生，但也有不少探索的乐趣。房子就是这样，你住多久，就有多久在调整。"他常常运来旧日的收藏品，放着看了感觉不对，又搬走，过几天再搬另外一堆东西过来尝试……"自己家当然一定要做出自己最喜欢的风格，新华路的家是10年前我最爱的风格：前卫时尚，极简主义，强烈的黑白对比；随着年龄的增长，现在更喜欢中性以及多元化的风格。看起来我们是花了5年建造这个家，其实这里的很多东西是花了十几年的时间慢慢积累而来的。"

设计师还有一样心头至爱，一套一九三几年出版的原版《丁丁历险记》漫画，是他花了多年时间在法国的二手市场上一本本搜集来的，始终收藏在身边，有时候拿过来随手一翻，闻到熟悉的味道就可以安心睡去。"什么时候我把它搬过来，"他笑着说，"就是完全把这里当成家了。

罗海峰 / Ocean Luo

奥迅室内设计有限公司 创始人、董事

罗海峰，从事室内设计工作15年以上，其擅长将生活的趣味元素与想像结合，以现代主义的简约风格为基调，在设计上做到空间体验最大化，打造能传递艺术和文化真实美感相结合的精品空间。曾获年度十佳设计师，年度公寓最佳设计师之荣誉。

2013年创立个人品牌奥迅设计，专注高端专业地产设计领域，以务实、专注、赤诚的态度服务于客户，专注地产和商业公共空间的策划设计与艺术陈设，在罗海峰的带领下，团队凭借丰富的设计经验以及崭新思潮，不断屡创独具之作。奥迅关注设计过程中的每一步，大胆地创造灵感，悉心地创建灵魂，用艺术的情怀再次传递设计生活的幸福，力求创造完美体验空间直抵人之心灵所需，经过三年的发展，设计团队规模逾百。

静享悠然生活
QUIETLY ENJOYING LEISURE LIFE

项目名称 | 广州星汇金沙花园别墅
设计公司 | 奥迅室内设计有限公司
设 计 师 | 罗海峰
项目地点 | 广东广州
项目面积 | 240 m²
摄 影 师 | FOCUS摄影工作室
主要材料 | 橡木饰面、英伦玉石、圣罗兰石等

DESIGN CONCEPT | 设计理念

Seeking for a peaceful land in the hustle and bustle city, quietly enjoying life and releasing the body and mind are the inner thoughts of modern people. This project belongs to Guangzhou Xinghui Jinsha Garden Villa. The designer sets concise and lively "Oriental elegant" style as the tone. The choices of modeling, materials and furnishings set off tranquil and elegant atmosphere of the space, collocating with light gray leather surfaces, which makes the elegant atmosphere more attractive. The unified space has some changes, making a contrast in the harmony, which stretches and relieves bodies and minds of the residents.

在喧嚣繁杂的闹市中寻获一方净土，静静感受生活，好好释放身心，是许多都市人藏在心里的念想。本案的主体是广州星汇金沙花园别墅，设计师以简约明快的"东方雅致"风格为主调，从造型、材料以及饰品的选择上，衬出空间静心雅致的氛围，再搭配浅灰色扪皮饰面等局部的点缀，使得雅致的氛围更加沁入人心。同时空间在统一中带有些许变化，和谐中产生对比的效应，使居住者的身心在这里得到舒展与释然。

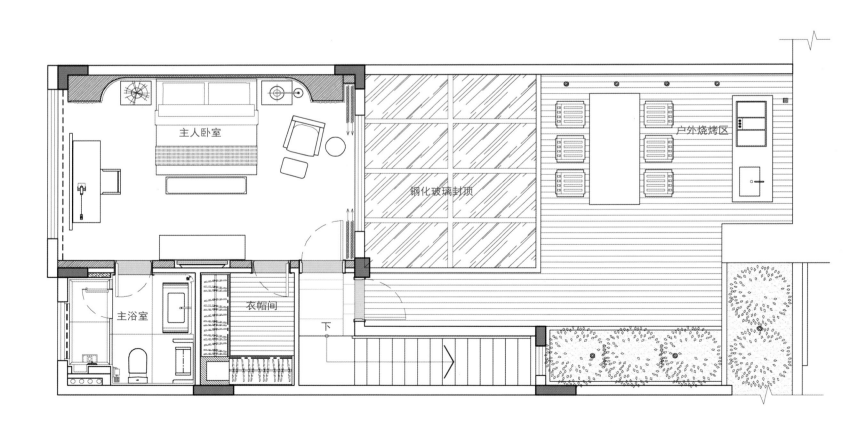

桂峥嵘 / Grace Kwai

上海桂睿诗建筑设计咨询有限公司　创办人兼艺术总监

桂峥嵘，INTERIOR DESIGN美国室内杂志十大封面人物，LUX莱斯室内设计杂志十大封面人物，上海交通大学海外学院艺术品鉴赏收藏与投资高级研究员等。2004年创立了上海桂睿诗建筑设计咨询有限公司。有丰富的项目设计经验，而且是专业的设计管理者。一直以尊重设计为出发点，强调学会尊重自己，才能让别人尊重你。不断突破、创新做出让自己喜欢，同时也受大家喜爱的作品。

2016年Asia Interior Design Award 设计银奖，2016年中国国际建筑及室内设计节"金外滩"最佳居住空间奖，以及2016年亚太室内精英邀请赛银奖；实力荣获了2014-2015年ID杂志中国室内设计年度封面人物；2014年度国际室内设计大奖"艾特奖"最佳展示空间设计、最佳别墅豪宅设计入围奖；2014年CIDA中国室内设计大奖"居住空间、样板间设计提名奖"和"公共空间、商业空间奖"等。

塞纳河畔的柔情
TENDERNESS OF THE SEINE RIVERSIDE

项目名称 | 南通上林苑别墅
设计公司 | 上海桂睿诗建筑设计咨询有限公司
设 计 师 | 桂峥嵘
硬装设计 | 李成祥
软装设计 | 张艳玲
项目地点 | 江苏南通
项目面积 | 510 m²
摄 影 师 | 钱达
主要材料 | 雕刻白大理石、碳化木地板、水晶棱镜、进口壁纸等

DESIGN CONCEPT | 设计理念

In the early morning after the rain, when we are walking on the Seine riverside to overlook the Notre Dame de Paris bridge, the ancient riverside brings us inspirations. With reference to the azure after raining and the Seine, we bring the charms of the Seine and the elegance of the azure into this villa show flat, combining with romantic French decoration style to try to explore a mysterious beauty.

雨后的清晨，当我们漫步在塞纳河畔远眺巴黎圣母院桥，古老的河畔带给我们启发。如今我们将雨后的天青色与塞纳河为引子，将塞纳河的妩媚迷人和天青色的典雅带入这套别墅样板房作品中，结合浪漫法式装饰风格，试图探索一种神秘的美感。

White clapboard and concise ceiling modeling match with white gray marbles, interspersing with azure, which outlines an as mysterious and elegant interior environment as the orchid in the valley. The shining and fashionable diamond cutting elements and diamond cutting model mirror collocate with light screening materials, which creates an as dazzling visual feeling as the jewelry and foils the gorgeous home. In the pure blue tone of the interior, we seem to appreciate the tenderness of the Seine riverside.

　　白色护墙板，简洁的吊顶造型，配合白灰色大理石，点缀以天青色，勾勒出一个如深谷幽兰一般神秘优雅的室内环境。辅以闪亮时尚的钻石切割元素，钻石切割造型的镜面，搭配具有丰富遮光感的材料，搭配出如同珠宝般令人目眩神迷的视觉感受，烘托出绚丽的家。在室内那抹纯净的蓝调中，我们仿佛领略到了塞纳河畔的柔情。

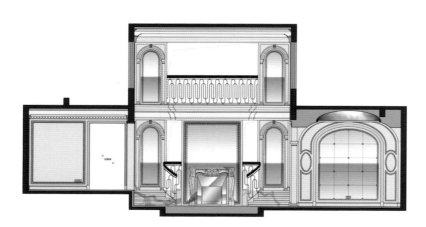
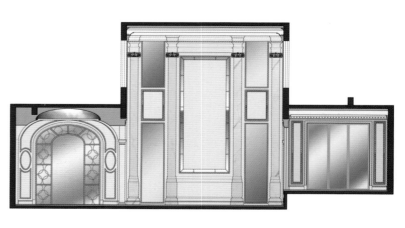

上海全筑建筑装饰设计有限公司 /
Shanghai Trendzone Construction Decoration Design Co., Ltd

trendzône　DecorGroup
全　築　装　饰

　　全筑装饰集团成立于1998年，是集建筑装饰研发与设计、施工、家具生产、装饰配套和建筑科技为一体的大型装饰集团，是中国建筑装饰协会常务理事单位和上海市装饰装修行业协会副会长单位，具有设计甲级和施工一级资质，是中国驰名商标和上海市著名商标企业。蝉联十二届上海市室内设计大赛金奖，在中国装饰业界极具实力及美誉度。

　　全筑装饰集团拥有专业化部品加工基地及先进的生产设备，为装饰装修提供强大保障。同时建有专门装饰配套展厅，为装饰装修提供全面、专业化的后期配套服务，并在行业内率先成立了自主研发部门，为客户提供更为专业的系统服务。

天人合一 精神内敛
ONENESS OF MAN AND NATURE, RESTRAINED SPIRITS

项目名称 ｜ 运河上的院子
设计公司 ｜ 上海全筑建筑装饰集团股份有限公司
软装配套 ｜ 上海全品室内装饰配套工程有限公司
项目地点 ｜ 北京

DESIGN CONCEPT ｜ 设计理念

This project is located in the east starting of the Chang'an Avenue on the riverside of The Grand Canal from Beijing to Hangzhou. Since the completion of the courtyard, it has been well praised in this industry. The accurate design concept, the reasonable space layout, the exquisite detail consideration and the cozy atmosphere leave deep impressions on people. The train of thought of the landscape planning is neo-Chinese landscape system formed by quiet street, deep alley, warm yard, floral stream and landscape garden. And logically the interior decoration adheres to the neo-Chinese style.

北京泰禾运河上的院子位于长安街东起点,坐落于京杭大运河河畔。院子自完工以来,在业内深受好评,准确的设计理念,合理的空间布局,精致的细节推敲,惬意的气氛营造,都给人留下深刻的印象。其景观规划的思路是由静街、深巷、馨院、花溪、山水园一起形成的新中式景观体系,而室内装饰也顺理成章以新中式作为依归。

Chinese courtyard stresses closed surroundings with spirit in the center, contacts with the spirits from heaven and earth and pursues "oneness of man and nature". This is the spiritual position of the courtyard. The jade symbol on the screen wall seems to be such a symbol. The new of neo-Chinese style means that its essence is still a contemporary design. So when Chinese elements are presented, the classical concepts are not old and form a perfect combination of their tensions and ear spirits.

中国院子讲究四方围合，中间藏气，与天地精神相往来，追求"天人合一"，这是中国院子灵性层面的定位，照壁上的玉璧符号或许是这样的一个象征。新中式之新意味着其本质还是一种当代设计，所以在中国元素得以呈现的同时，那些古典的概念并不显得陈旧，而是将自己的张力与时代精神形成了一种完美的结合。

亚太名家荟萃·风靡大陆 | 173

亚太名家荟萃·风靡大陆 | 177

王小根 / Xiaogen Wang

北京根尚国际空间设计有限公司 总经理、设计总监

　　王小根，根尚国际设计机构创始人兼设计总监，IAI亚太设计师联盟理事。在室内高端地产设计领域拥有近20年的丰富经验，"以刚毅而柔美的设计理念，创造一种极致的生活文化"为追求。自2006年创立根尚国际至今，曾荣获中国十大样板房设计师、亚太室内设计样板房设计大奖；代表作品8哩岛曾荣登台湾奢华杂志ARCH《雅砌》封面，并荣获2015年度美国《莱斯》和美国《室内设计》杂志中文版年度封面人物；被中国建筑装饰协会评选为2015年度中国设计青年领袖；2016年当选为中国建筑装饰协会设计委员会副主任委员。

天青处
THE AZURE PLACE

项目名称 | 淮安生态新城建华观园别墅
设计公司 | 北京根尚国际空间设计有限公司
主持设计师 | 王小根
参与设计师 | 王树宇、吴霞等
项目地点 | 江苏淮安
项目面积 | 430 m²
摄 影 师 | 金选民
主要材料 | 紫檀木、青瓷色混油木作、法国紫彩石材、金丝白玉石材、仿古铜等

DESIGN CONCEPT | 设计理念

Based on Eastern traditional culture, the whole space uses modern elements to make a perfect comparable and conflict encounter and creates a modern living residence with old Oriental cultural connotations.

Zhao Jie, the Emperor Huizong of Song Dynasty, described the color in his dream as "the color when rain stops, sky is azure and the cloud appears". Inspired by this, the space conveys Zen-like connotations that the water and sky merge in one color and the dynamic and static states integrate. The "azure" of Ru Kiln is a very special color of porcelain. The best time it fires is in misty rain. The azure misty rain is beautiful, tranquil, deep and clean. The entire space uses azure as the main color to deduce the gradually clear natural beautiful scenery after raining. At the same time, the red sandalwood color and beige create tranquil atmosphere. The space is interspersed with blue and orange in small area, which activates the space and adds a sense of rhythm.

整体空间根植于东方传统文化之上，以现代元素为主，让对比与冲突完美邂逅，打造了一个具有古老东方文化底蕴的现代风格居所。

以宋徽宗赵佶为汝窑钦定色时的一句"雨过天青云破处，这般颜色做将来"为灵感，在空间中传递出水天一色、动静如一的禅宗意蕴。汝窑的"天青色"，是瓷器中很珍贵的一种颜色，烧制的最好的时光，是在烟雨天气中。天青色等烟雨，美丽静谧，深邃清澈。整体空间以天青色为主色，演绎了雨后天空逐渐放晴的自然美景。同时，辅以紫檀、米色的使用营造出静谧气息，空间在小区域点缀了蓝色与橙色，活跃空间，增添了律动感。

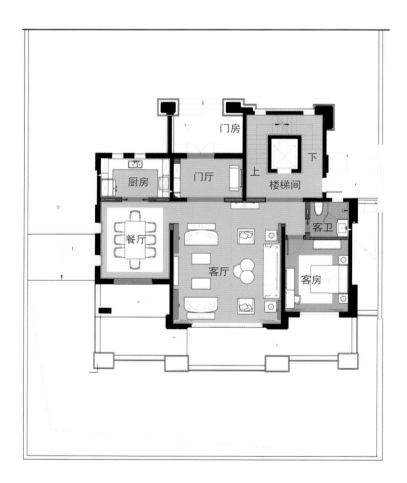

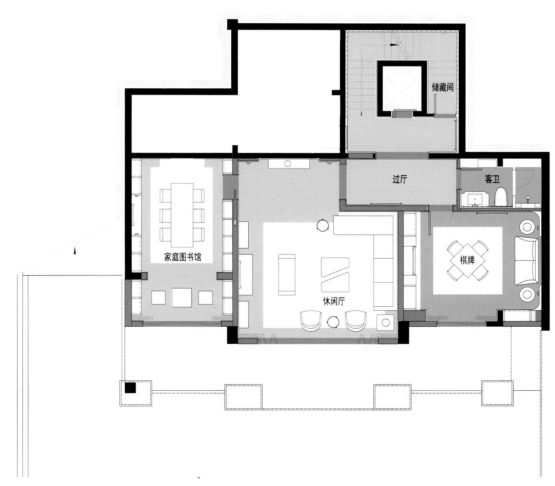

The ceiling of the public area is a "water drop" modeling made of transparent mirror stainless steel as if the last drop of water after raining spreads over the quiet water surface.

The wall lamp of the living room is white marble water drop modeling. The carpet is water ripple pattern. The intersection of water drop elements presents poetic Oriental beauty.

公共区域顶面天花是一个通透质感的镜面不锈钢材质的"水滴"造型,犹如雨后最后一滴水珠,在平静的水面上荡漾开来。

客厅壁灯为白色云石水滴造型,地毯为水波纹涟漪图案,水滴元素相互交织,展现出诗意东方的美。

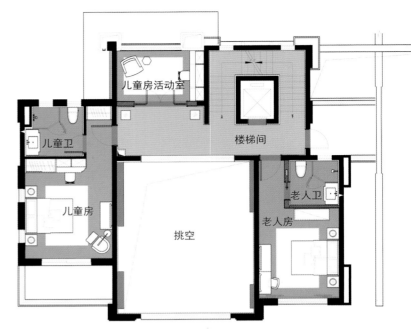

Designs of other spaces continue the theme space. In the first floor underground, white birch trees in the painting, branch pattern carpet in family library and the design of water cluster glass droplight present natural images of azure sky, dynamic water and tranquil forest. The ink splash pattern carpet, round water drop glass droplight and vertical wood grid crafts of the ceiling in the master bedroom create vivid flowing sense of the space.

其他区域的设计同样是主题空间的延续，负一层画品中白桦树林、家庭图书馆的枝桠图案地毯、簇状水流玻璃吊灯设计展现了天青、水动、林静的自然意向；主卧室的泼墨图案地毯、圆形水珠玻璃吊灯、吊顶的竖型木格栅工艺，同样营造了气韵生动的空间流动感。

亚太名家荟萃·风靡大陆 | 187

李益中 / Yizhong Li

李益中空间设计（深圳、成都） 都市上逸住宅设计 创始人、设计总监

　　李益中，大连理工大学建筑系学士、意大利米兰理工大学设计管理硕士、深圳大学艺术学院客座教授及中国建筑学会（全国）理事。2010年受聘为清华美院、中央美院建筑学院等艺术院校的设计实践导师，2012年受聘深圳大学艺术学院客座教授。

　　2015-2016 年获年度杰出设计师、年度杰出室内设计奖年度杰出设计企业、中国室内设计年度封面人物、2015 年获中国室内设计年度最佳别墅设计、中国家居时代精英设计师奖、BEST 100中国最佳设计、中国家居时代精英设计师奖。2013年获金堂奖中国十佳酒店设计奖，中国室内装饰协会全国室内装饰优秀机构。2007 年出版著作《样板生活》，2009年出版《FROM B TO A 售楼处设计策略》。

少即是多
LESS IS MORE

项目名称 ｜ 东莞鼎峰源著
设计公司 ｜ 李益中空间设计
硬装设计 ｜ 李益中、范宜华、段周尧
软装设计 ｜ 熊灿、欧雪婷、欧阳丽珍
项目地点 ｜ 广东东莞
摄影师 ｜ 郑小斌
主要材料 ｜ 木地板、布鲁塞尔木纹、白金沙、木饰面、布艺、墙纸等

DESIGN CONCEPT ｜ 设计理念

The characteristic of modern space design is conciseness and its essence lies in the flowing of space. Earlier in the past, because of the limitation of construction technology, the interior space partition of architecture is rigid and closed. Until the rise of modern architecture movement in the beginning of the 20th century, because of the reformation of construction materials and technology, space is freed from the limitation of structure and achieves flowing sense.

现代空间设计的特点是简洁，而其精髓更在于空间的流动。更早以前，因为建筑技术的限制，建筑内部的空间分割是刻板而封闭的，直到二十世纪之初的现代建筑运动的兴起，由于建筑材料及技术的革新，空间才从结构的桎梏当中获得自由，实现了空间的流动。

People need shelters to keep out wind and rain and keep safe. But they need freedom beyond limitation and imprisonment of the space structure and become the owner of the space.

Freedom and flowing of the space is essential for human nature. Only when one's body is not limited by the space can his sprit be easier to stretch. This kind of space is full of energy and has positive effects on one's mind. Free and flowing space is full of sequence and rhythm with changeable depth of fields, so it is easier to produce poetry.

Creating freedom and flowing of the space is the origin of our creation.

人需要界面的庇护，以遮风挡雨，获得安全；但人又必须超越空间结构的限制与禁锢，获得自由，成为空间的主人。

空间的自由与流动，是人性的必需。只有当人的躯体不被空间区隔禁锢，人的精神才更容易获得舒展。这样的空间是富有能量的，对人的身心施以积极的影响。自由流动的空间，充满序列、节奏，有景深层次的变化，因而也更容易产生诗意。

创造空间的流动与自由一直是我们创作的原点。

This project has two houses with two elevators which are exclusive. So the elevator spaces can be used as a porch and the transformation of entering the house.

The living room and dining room are laterally set and designed, because the variable tea room partition and transmission design of kitchen, bar and dining room achieve longitudinal stretch of the space. Living room, tea room, recreational balcony, dining table, kitchen and bar share the living space and gain freedom and flowing. Various life activities happen in this space, which improves considerations and communications between family members and creates harmonious family atmosphere.

本案为两梯两户，电梯专属，因而电梯空间可以当作玄关空间使用，完成入户前的空间转换。

司空见惯的客餐厅横向并置设计，因为茶室空间隔断的可变性设置，以及厨房、酒吧及餐厅的穿透式设计，实现了空间纵深方向的拓展。由客厅、茶室、休闲阳台、餐厅、厨房、酒吧等功能模块共享串连的生活起居空间，获得了自由与流动。各种生活活动在这个空间里共享发生，促进了家庭成员之间的相互关照、相互交流，利于形成融洽的家庭氛围。

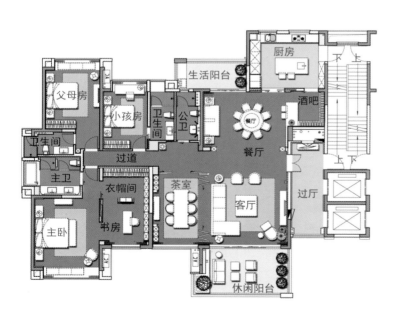

For private bedrooms, they are placed at the end of the space sequence, are connected by corridors, are independent with their own bathrooms and have their own world. The master bedroom is equipped with bathroom, cloakroom and small study, which is partitioned yet not disconnected and creates a small flowing of the space.

The fundamental purpose of interior design is to create space not to decorate the surfaces. When the space pattern has produced rich rhythmic changes, we advocate moderate and appropriate surface designs.

For Ludwig Mies Van der Rohe's saying "Less is more", most people cannot understand its real meaning. It is widely understood as concise forms and exquisite details. But in fact its essence is creating "more" space changes by "less" forms.

Less is more. There is a mind guiding my design practices for several years, that is how to use conciseness to perform richness. In addition to the pursuit of creating changeable spaces, materials, colors, lights including natural lights and artificial lights, how to display these design elements and how to collocate them can create changeable and rich layering and elegant life beauty.

In this project, in order to adopt the structure of the space, we try to use concise forms to create modern Eastern meanings. East means intangible and elegant, implicit and reserved. The space is flowing, which doesn't mean you can get everything at one glance. The multi-levels and changeable treatments of the space offer Oriental experiences of different views in different places. Oriental cultural attributes are achieved by relatively light scales, linear expressions of forms, symbolizations of furnishings and artistic conceptions of decorative paintings. Space, surface and furnishing complement to each other and create Oriental connotations.

对于要求私密的卧室空间来说，被置于空间序列的末端，由走廊连接，各自独立，各自配套卫生间，成为自己的小天地。主卧室恰当安排卫生间、衣帽间、小书房的面积配比，隔而不断，创造空间的小流动。

室内设计的根本目的在于创造空间，而不是为了修饰界面。当空间的格局已然产生丰富的节奏变化，我们推崇节制而到位的界面设计。

　　密斯的名言"less is more",大部分人并不能理解其真正的含义,普遍理解为简约形式加上其精巧的细节,而实际上其本质是,用形式的"less"创造空间变化的"more"。

　　less可以很more。我这几年的设计实践一直有一个思想在指导,就是如何用简洁去表现丰富?除了去追求本质营造多变空间,材质、色彩以及光(包括自然光、人工光),还有陈设这些设计要素,以及它们之间的搭配处理,都可以创造多变而丰富的层次,创造高雅的生活之美。

　　在这个案子里,因应空间的架构试图以简洁的形式去营造现代东方的意味。东方意味着空灵飘逸,意味着含而不露。空间是流动的,但并不代表一览无遗。空间的多层次可变性处理,给予空间步移景换的东方式体验。东方的文化属性同时以相对轻盈的尺度,形体的线性表达,陈设的符号化,以及装饰画的意境表达来实现。空间与界面、陈设等要素相辅相成,完成东方意蕴的塑造。

连自成 / J.K Lien

大观·自成国际空间设计 设计总监 英国De Montfort大学 设计管理硕士

连自成先生曾游学英国，深受欧洲深邃的文化生活的熏陶，为设计的创意思源和风格奠定了一定的基础。之后转至上海，继续执着于室内设计的行业，屡创精彩的设计佳作。

Mr.连多年的室内设计经历，以精品ELITE、建筑感、艺术性、客制化为设计的着准点，强调文化、品质的完美融合，倡导更优质的生活设计态度。从创立至今，已在五星级酒店、豪宅、会所、样板房、企业总部、商业空间等中树立了良好口碑。

曾获得Asia Pacific Interior Design Biennial Awards；《王子晶品》荣获上海十大明星楼盘第一；《上海日航酒店》获亚太APIDA酒店设计银奖；2012年荣获第十届国际传媒奖年度杰出设计师奖；2013年，当选了由美国《室内设计》中文版及CIID主办的"2013-2014中国室内设计年度封面人物"。2015年其作品入围日本Good Design优良设计，获得英国FX International Awards 提名，此外两次荣获北美著名设计大奖"GRANDS PRIX DU DESIGN"，同时，入选2015Pchouse时尚设计盛典中国十大高端室内设计师。2016年分别获得意大利国际大奖A'design award 银奖。2016年作为"心+设计"学社社长与十一位心社设计师受邀参与国际性设计展览米兰三年展Triennale "Design after design" 等。

海派密林里的新中式
A NEO-CHINESE STYLE IN VARIOUS SHANGHAI STYLES

项目名称｜海珀黄埔
设计公司｜大观·自成国际空间设计
设 计 师｜连自成
项目地点｜上海
主要材料｜大理石、不锈钢、布艺等

DESIGN CONCEPT ｜ 设计理念

Through interpretation of neo-Chinese style, the designer extracts elements and uses them in house type design. First starting from the layout, the challenge is how to create more possibilities of the space and how to control the design "scale" in residents view. Though the later sounds very abstract, it needs designer's accurate calculations, such as collecting amount in storage space, capacity of the kitchen and proportion of floor height and furniture. The scale of space kinetonema is well arranged to squeeze more reasonable areas from other spaces for a certain function space. This is where the difficulty lies.

通过对新中式的诠释，从设计中提取元素将以运用到户型设计中。首先从布局出发，其考验在于如何挖掘空间的多种可能性，以及通过居住者的视角把握设计上的"尺度"。尽管后者听上去很抽象，但却需要设计师精准的计算，例如：储藏空间中收纳的量、厨房的容量、层高与家具的配比等。空间动线的尺度，一分一毫妥善安排，为了某个区域功能要从其他地方挤出更多合理的空间，这就是难度所在。

As a place to display collections in the porch, the antique-and-curio shelves inject Chinese culture and art into the space without conflict, which reflects life trends of Chinese in an international vision. The sliding doors and partition between the dining room and kitchen skillfully control the space, which increases penetrability by movable partition. The side table and furnishings near dining table and lamps with Chinese lines foil the Chinese aesthetics. In addition, neat decorative lines and sedate living colors echo with each other. We can also see Chinese furnishing articles in a casual corner in bedroom, which outlines cultural flavors of the interior.

巧妙地将博古架作为一处玄关陈列藏品，将中式的文化和艺术置入空间而不突兀，以国际化的视野来审视中国人的生活走向。餐厅与厨房之间的移门与隔断，精妙地将空间进行操控，可合可开的分割令空间增加了穿透性。餐桌一侧的边桌与装饰品，中式线条的灯具，均衬托出了新中式风格的美学。除此之外，硬朗的装饰线条与沉稳的居室用色相呼应，我们还可以在卧室等不经意的角落看到中式摆件的应用，勾勒出室内的文化气息。

席卷港台

SWEEPING HONGKONG AND TAIWAN

谭精忠 / David Tan

TAD大隐设计集团 董事长

生于台湾，资深当代室内设计师、艺术收藏、跨界策展实践者。1989年，成立谭精忠室内设计工作室，1999年，相继成立动象国际室内装修有限公司（台北）、大隐设计集团（上海）、设艺之间、君远建筑师事务所、大译国际室内装修有限公司。

设计风格以"现代再新东方"为标志，将当代艺术与空间结合，东方元素的应用在业内独树一帜。作品多次荣获国内外室内设计大奖；连续多年入选《中国环境设计年鉴》、中国十大当红设计师等。任教台湾实践大学讲师、清华大学环境艺术学院讲师。

自1999年起，涉足艺术、空间跨界策展。先后举办：《一瞬之光》《无声之诗》《总统的秘密》《有龙则灵》《野人献曝》《体·验》等展览。

艺玩·质朴
ART INNOVATION AND PURELY INSPIRED

项目名称｜西安中大项目4#楼26层D户样板房
设计公司｜动象国际室内装修有限公司
设 计 师｜谭精忠
项目地点｜陕西西安
主要材料｜实木复合木地板、镀钛不锈钢、米黄洞石、皮革、玻璃马赛克、绷布板、夹纱玻璃等

DESIGN CONCEPT ｜ 设计理念

This project nestled in the fancy area in Xi'an where represent the epitome of big city. This model house is located in the twenty-sixth floor, and it offers panoramic views of surrounding landscape which highlights the value of this project.

本案位处于西安市高新区，高端城市综合体之中，拥有顶级地段及绝佳景观视野。此样板房坐落于26楼，设计上借助落地整面窗扇导入自然景观，成就豪宅视野的宽泛及大气，突显本案的价值。

Art innovation and purely inspired are the main concepts of overall design. Art innovation presents for life attitude of high educated and elegant taste group of people. In addition, the designer's choice of high quality furnishing and artwork further enhance the sense of comfort within this space.

艺玩·质朴是本案设计发想的主轴。艺玩代表文人雅士的生活态度，质朴则是以纯粹的、肌理丰富的木质素材来表现雅痞生活的低调品味，艺术品巧妙陈设显现出空间所隐喻的生活态度与品位，点缀性的融入不锈钢镀钛与灰镜等材料，为空间增添了时尚感。

After walking through the door of a building, you'll see the entrance is composed of a constituted cabinet that establishes a complex relationship between wired glass and titanium metal. This design can partly penetrate and extend to the multi-functional space, where the skillful collocation of materials balances the heavy tone of the space. There is a contemporary art sculpture in the middle of the porch, which sets the sedate and restrained tone of the space. The wallboard provides better places for clothes, shoes and hats with perfect unified functions and spatial view.

The living room and dining room are on the two sides of the porch respectively. The living room is open with nine meters high. The space is wide open, which provides kind of high visual impact. The flat walls are covered with special paint. The hidden modulator tube on the top titanium plate of the dining room presents distinct visual flavors. The exquisite details make the space full of tension and sense of layering, which is solid and inclusive.

玄关开启入户大门，映入眼帘是以夹纱玻璃及镀钛金属结合之端景柜，视觉可局部穿透延伸至多功能空间，质材的巧妙搭配平衡了空间较浓的色调。在玄关中间处置放当代艺术雕塑品，奠定空间沉稳、内敛的基调。壁板造型结合收纳、衣物、鞋帽各置其所，功能使用与空间视觉感完美统一。

玄关左右两边分别是客、餐厅。客厅区为开放式空间，宽约9米，视野空间开阔，空间气度尽显。壁面运用了特殊漆，延续至各个空间。餐厅顶面镀钛板内暗藏灯管带来独特的视觉韵味，精致的细节处理铺陈出空间的张力与层次感，稳健而具包容。

The multi-functional space is between the porch and living room. In order to make visitors feel relax and comfortable while entering the space, the designer presents a private garden space by using elegant tall cabinet and light. The interspersing of droplights creates the atmosphere of tasting wine. It provides a place of flower arrangement, afternoon tea and chatting for the hostess. The heart can be comfort and then the temperament and interest can be freely achieved.

多功能空间位于玄关与客厅之间，运用造型精致的高柜，淡化了空间压迫感，宛如室内花园凉亭。在吊灯的点缀下，既可打造出品酒的娱乐所需，亦可为女主人插花、午茶聊天的小天地，心灵得以慰藉、情趣即可实现任由发挥。

The dining room was designed for privately customized reception center. The exquisite and fashionable furniture and the collected art works present the taste of beauty. The main design presents the high value of this residential building and the great taste of the residents.

"Light meal kitchen" is the main concept design. The designer makes the functional segmentation of the kitchen more obvious and the entire space is more wide open. The ceiling lighting is made of delicate pieces of assembled & wired glass. The soft light is the main design concept for this space. The designer believes that the chef would enjoy cooking while the warmness is around the space.

餐厅空间主体呈现私人会所宴客的情境，精致时尚的家具与典藏的艺术品相互辉映，生活美学与空间艺术自由对话，均描绘出统一又富有层次的画面。

"轻食厨房"中岛的轴线设计将厨房机能分割更加明显，整体空间设计流畅。天花板照明以夹纱玻璃材质进行美化，柔和的灯光搭配一应俱全的料理设备，使得生活场景一一映现，主厨在此乐享其中。

The main ideal design for the master bedroom is the warmness and comfort. The headboard matches leather with large area of steel brush wood veneer, extending from ceiling to cabinet, which creates a complete collecting sense of the space. The dress room uses the same brush wood veneer to collocate with fabric background, which improves the delicacy of the space. The most unique spot light for this design is that the open design is good for the whole space partition and position. The surface of the cabinet adopts small scale cabinet by reflecting the elegant taste of the owner. Passing through the dress room and entering into the master bathroom, the dark color stone floor and surface foil the high-texture equipments. The independent bath leans close to the French window, which achieves all-round and comfortable feelings when the owner is bathing.

主卧室的设计以温馨、舒适为主调，床头板以皮革搭配大面积钢刷木皮，从天花延伸至机柜，营造出空间的完整容纳感。更衣间同样以钢刷木皮的框架搭配壁布背板基底提升了空间的精致度。而开放式的设计善用了整个空间分割及定位，柜内采用贴心的小规划机映衬主人井井有序的生活模式。经过更衣室进入主浴室，深色的石材地坪及台面衬托着高质感的主卫设备，独立浴缸依偎着落地窗实现沐浴时独享360度的全方位舒适感受。

The design for elder's room is bright and comfortable with clean and neat lines. The headboard uses large area of leather, and the floors are covered with handmade carpets, which is exquisite and elegant. The contemporary art sculptures add additional value. Under the strong urban leisure and self-enjoyable atmosphere, the starting point of the bathroom is the main function of design. Large area of glass window leads into the outside scenery. The main ideal of the design is combination of clients' ideal life style and real time interaction.

The purpose of living room is pursuing the best quality of life which presents the average of people, things, objects and conditions. The main wall continues to use the same stone as in the multi-functional space, collocating with wired glass door, which produces visual extension and becomes common languages in the space. Large area of glass window brings into the outside scenery, which is quiet and restrained. It creates a complete combination of static and dynamic.

孝亲房空间明亮、舒适，整体线条干净利落。床头板使用了大面积的皮革，地面铺设订制手工地毯，精致雅丽，当代艺术雕塑品的置放更增添了一份韵味。而浴室在强烈都会休闲和自我享受中仍以机能作为出发点，大面玻璃窗眺望户外景致，充分的光线与干湿分离的规划，在显出此区的沉稳与实用性，并将住家养身的安全贯彻于此空间。

起居室以寻求人、事、物、境最佳平均质为设计方向，让所有事物能各安其所，安静而有质地。因此主墙延续多功能空间的石材，搭配夹纱玻璃门产生视觉延伸性成为空间中的共同语汇。大面玻璃窗眺望户外景致，安静不喧哗的内敛质感，塑造了完整的动静皆宜——自在、真实而贴近人性的设计实现居家品质的相互结合。

亚太名家荟萃·席卷港台 | 223

郑树芬 / Simon Chong

SCD（香港）郑树芬设计事务所 创始人、设计总监

郑树芬，香港著名设计师，英国诺丁汉大学硕士，"雅奢主张"开创者，主张奢侈以"雅"为度的设计理念。他长期置身于中西方文化的研究，被媒体誉为亚洲最能将中西文化融入当代设计的香港设计师，极其强调项目的文化内涵，用"时尚典雅、内敛惊艳"的设计手法完成了诸多社会名流、明星的豪宅官邸。

90年代开始进入中国内地市场，专职为房地产、私宅高级定制及高端商业空间服务，因此有业内人士为其作诗"名门豪宅百胜开，内敛惊艳炫高台。绅士风度心如月，原创明师笑将来"。其设计项目遍布中国、日本、东南亚、欧洲等地，在业界享有良好的信誉和口碑，设计作品屡获设计界殊荣，同时受到国内外媒体的广泛关注，自2004年起发行自己的作品集《居室韵律》、2006年《构》、2016年《雅奢主张》。

雅奢主张：文化无界
ADVOCATING ELEGANT LUXURY, CULTURE WITHOUT BOUNDARY

项目名称｜香港深水湾·文礼苑
设计公司｜SCD（香港）郑树芬设计事务所
主案设计师｜郑树芬
项目地点｜香港
项目面积｜500 m²
主要材料｜木地板、墙纸、布料、瓷砖、皮革、玻璃等

DESIGN CONCEPT ｜ 设计理念

This is a home of a foreign family of four who want it relaxing, natural and warm! There is always sunshine in the warm home, moreover there are beautiful sceneries of green hill and blue sea to appreciate outside the window. So Mr. Chong uses glasses in many spaces such as dining partition and glass wall in the living room to bring outside beautiful scenery into inside. The beautiful sceneries of green hill and blue sea outside the window are leaded into inside naturally under the sunshine. It uses travertine stone, wood floor with natural texture and wall cloth. The displaying scene is natural, the relaxation inside which belongs to the freedom of home.

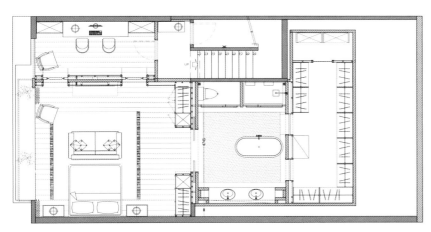

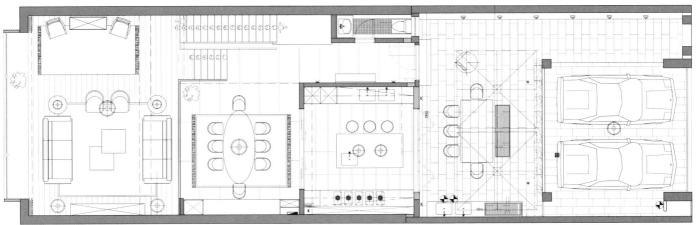

这是一个外籍人士的四口之家，客户希望深水湾这座居所轻松、自然，有家的温度！温暖的家总是有阳光，何况窗外还有青山蓝海的美景可欣赏，于是郑先生在空间多处使用了玻璃，餐厅隔断、客厅玻璃墙等，将室外的自然美景引入室内。窗外青山蓝海的美景，在阳光照耀下自然引入居室，空间使用了天然洞石、自然肌理感的木地板、墙布，陈设场景自然不做作，置身其内的放松，应该是属于家的自由。

Many foreigners like to invite Mr. Chong to design their home, which is because they are attracted by the classic and unbounded combination of Chinese and Western cultures by Mr. Chong, and the owner of this home is no exception. They love their own culture and yarn for the implicit and mysterious Eastern civilization. There is no doubt that Mr. Chong can give them the satisfied answers!

The design proportion of hard decoration is concise and refined. The materials express the combination of natural, plain and craft aesthetics perfectly, such as natural texture of travertine stone and wood, the combination of stainless steel and glass, matching with top luxurious brands in the world, American Baker furniture, Italy Promemoria furniture and so on.

The soft decoration refines and creates artistic atmosphere. In bigger aspect an artistic painting from auction room and in smaller aspect a pair of mandarin ducks article, all are authentic works by famous contemporary artists, meaning love and happiness and fully expressing the meaning of Oriental culture. The perfect combination of entire furniture texture and art works creates the real meaning of elegant luxury. At the same time the whole space makes an unbounded combination of Chinese and Western cultures.

很多外国客人喜欢找郑先生做设计，源自于他们对郑先生将中西文化经典无界结合的手法所吸引，这个家的主人也不例外。他们眷恋自己国家的文化，又向往东方文明的含蓄和神秘，郑先生无疑给予了他们满意的答案！

硬装空间设计比例简洁、精炼，材质表达将自然、质朴、工艺美感结合到位，如天然洞石、木头的自然肌理，不锈钢与玻璃的工艺组合，配以全球奢侈品牌：美国Baker家具、意大利Promemoria家具等顶级奢侈产品。

而软装方面则提炼和创造了艺术氛围，大到拍卖行的一幅艺术挂画，小到一对鸳鸯摆件，都是当代著名艺术家的真品，喻意爱和美好，全面表达东方文化意义。整体家具的质感与艺术品完美结合，缔造了雅奢真正的含义，同时整个空间将中西文化进行了无界结合。

Real art is always the most luxurious language of the space. The center of the living room is *Three Beauties* by Chinese artist Mr. Ding Xiongquan. The fresh and colorful picture makes you feel in spring, which is full of romantic and ebullient atmosphere. The vitality is obvious, making the whole space full of vitality.

The opposite wall is decorated with two paintings by Taiwanese artist Mr. Yao Ruizhong, *Good Time— Living in Seclusion* and *Good Time— The Autumn Sound*. The two art paintings are symmetrical and the color collision makes asymmetric beauty. The pictures are modeled after transformative landscape in late Ming dynasty. The single color matches with gold foil and beautiful landscape, which reflects the ideal reclusive life and reappears the elegant sentiment of ancient literati.

Every bedroom has a wide window, which provides ventilation, light and natural scenery. The family can luxuriously enjoy every moment of life in the comfortable space.

真正的艺术永远是空间最奢侈的语言。客厅正中央是中国艺术家丁雄泉先生的《红粉三美图》，色彩明艳的画面，宛如置身在春天一般，充满浪漫、热情洋溢的气息，饱满的生命力呼之欲出，让整个空间充满活力。

对面墙壁上挂的是台湾艺术家姚瑞中先生的《好时光-幽居图》《好时光——秋声赋》，两幅艺术画在空间的形式是对称的，而色彩的撞色则产生了非对称美。画面以晚明变形山水为摹本，于单色配金箔及奇山异水中投射大隐隐于市的理想生活，重现古代文人的高雅情操。

每一间卧房都有宽阔的飘窗，满足通风、光线与自然景致，家人们能在舒适的空间里，奢享生活中的每一寸光阴。

Real design is melted into life. The combination of sunshine, art and music makes them being together with their favorite things in life and gives them the most comfortable, desirable and sincere living enjoyment.

This is the Manderly Garden in Deepwater Bay in Hong Kong. The space is transparent and full of sunshine and natural flavor. Elegant and concise lines and the amazing art paintings are not sensory stimulations at first sight, but elegant designs which integrates real luxury into life.

真正的设计是融于生活的，阳光、艺术、音乐相结合，让他们在生活中与自己的喜爱之物朝夕相对，给予他们最舒适称心、最诚挚的居住享受。

这就是香港深水湾·文礼苑，通透空间，充满阳光与自然气息，优雅简洁的线条、画龙点睛的艺术画，不是第一眼惊艳感官的刺激，而是把真正的奢华融入生活中的优雅设计。

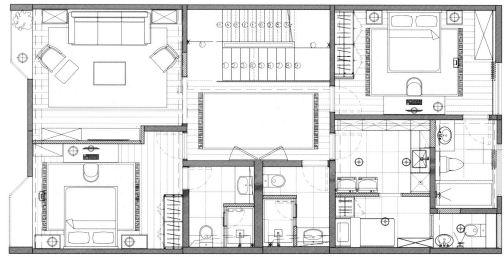

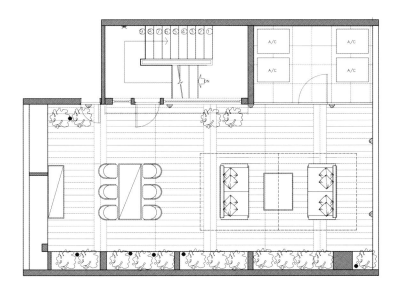

张清平 / Qingping Zhang

天坊室内计划有限公司 主持人

张清平,逢甲大学室内景观学系现任讲师。以品八分、养新气、融不同、立新意、传经典的设计理念,成就天坊今天的设计地位。坚持一个空间,不可能没有故事。哪怕只是短短的几句话,或是一个微小的感觉,不同的空间将带来不同的期望。对于蒙太奇的坚持,则让创造了一个属于天坊的空间语言。洋为中用,中为洋用。远看国际化,近赏中国化。

国际奖项的常客,台湾唯一连续5年入选为"英国,安德鲁·马丁室内设计年度大奖"华人50强、全球100大顶尖设计师、香港Perspective透视大奖、香港A&D Trophy Award、英国FX国际室内设计奖、美国IDA国际设计大奖、亚太设计双年大奖、台湾室内设计TID大奖、艾特国际空间设计奖Idea-Tops等。他拥有太多脍炙人口的作品,还不遗余力地向世界讲述着东方的故事,开创了不一样的新奢华——Montage(蒙太奇)美学风格!

2016年英国Andrew Martin安德马丁国际室内设计大奖,2016年德国reddot award红点设计大奖——Best of the Best最佳设计奖,2016年德国reddot award红点设计大奖——Red Dot红点奖,2016年意大利A'Design Award Competition,2016年英国SBID International Design Awards,2016年英国FX international interior design award,2016年英国LEAF Award,2016年日本JCD Design Award 商空大赏 BEST100,2016年法国INNODESIGN PRIZE国际设计大奖,2016年中国Jintang Prize金堂奖,2016年中国BEST100最佳设计大奖,2016年中国CIID室内设计大奖赛等。

中体西用的新东方艺术
NEW ORIENTAL ART OF WESTERNIZED CHINESE STYLE

设计公司｜天坊室内计划
设 计 师｜张清平
项目地点｜福建泉州
项目面积｜1672 m²

DESIGN CONCEPT ｜ 设计理念

The design concept of this project is summarized as westernized Chinese Style with round top and square ground, which is bright and spacious with real and visual intersections. It implements the design concept on the aspects of proportion, balance and circulation, refines meaningful totem elements from Oriental classics to deconstruct the essence of the style and rebuilds the space by intersections of real states and artistic conceptions, which renders meanings and characteristics of the space and presents strong opinions about westernized Chinese Style with round top and square ground. The integration of strong generalization ability and extreme conciseness strengthens the strong power produced by beauty and thoughts. The main axis is the combination of modern crafts and Oriental humanity art, which creates beauty of the entire picture and produces space tension as the film plot. Putting Western low luxury of Lounge into artistic conception of Oriental traditional life creates an endless and extensive luxurious space.

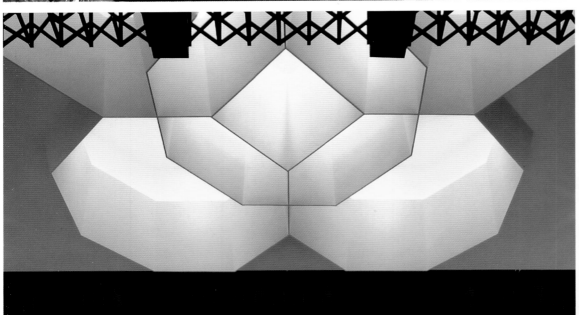

　　将本案的设计概念总结为：天圆地方中体西用，虚实交错明昶宽阔。从比例、均衡、行气三大方向落实设计概念，自东方经典中提炼出意义性图腾元素，解构风格本质，用实境与意境穿插，重新组构空间，渲染出空间的意义与特征，表现空间天圆地方中体西用的强烈主观，将巨大的概括力和极度简洁融合，强化美感和思想产生强烈的感染力。以现代工艺结合东方人文艺术为主轴，不仅架构出整体画面的美感也产生电影情节般的空间张力，将西方Lounge的低奢，剪切置入东方传统生活意境里，创造无限延伸的新奢华空间。

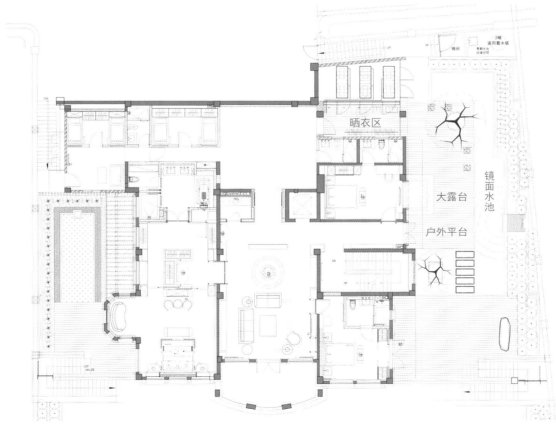

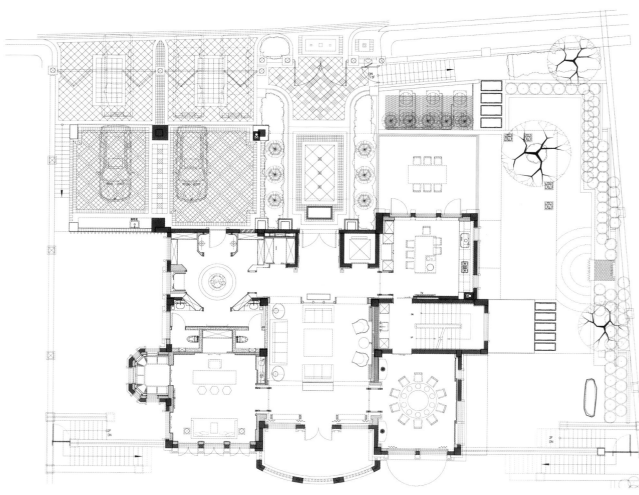

亚太名家荟萃·席卷港台 | 251

刘荣禄／Runglu Liu

咏义设计股份有限公司 执行董事、设计总监

刘荣禄，2002年成立咏义设计股份有限公司，2014~2015年期间担任亚洲大学室内设计系讲师，2016年担任亚洲大学室内设计系企业导师。2014年台北市立美术馆30周年典藏展计画的艺术场域，刘荣禄设计总监为首位以艺术家身份受邀参加这个空间艺术计画的设计师。

近年的获奖纪录有：2016年美国9th International Design Award (The convergence of light, shadow, and space)、2015年英国FX International Interior Design Awards 、Andrew Martin the international interior designer of the year award、2014年德国红点设计大奖产品设计类室内设计奖 、TID台湾室内设计大奖单层住宅设计奖、第十二届现代装饰国际传媒奖展示空间大奖及中国(深圳)国际室内文化节大中华区十佳设计师等多项荣誉。

几何的深度
DEPTH OF GEOMETRY

设计公司｜咏义设计股份有限公司
设 计 师｜刘荣禄、邱如怡、谢宜臻、黄沂腾
项目地点｜中国台湾
项目面积｜310 m²
摄 影 师｜郭家和
主要材料｜锈铜砖、银灰石、人文砖、木纹砖、大理石、茶镜、墨镜、明镜、橡木染咖啡、橡木染灰、橡木染白、橡木喷白、柚木深刻纹、黑铁烤漆、银玻璃、茶玻璃、绷布、实木条格栅、乳胶漆、进口壁纸等

DESIGN CONCEPT ｜ 设计理念

Details of implicit Dutch style design in the space echo with each other, which creates warm, magnificent and beautiful living environment.

The space has Dutch geometrical aesthetics yet not cold. The central sofa area plays an indispensable key role. It can be not only one direction as general sofas, but also multi-directions. However multi-function sofa is not complicated, it is the interior scenery people see when on the sofa, which determines the property of the sofa. Without aggressive construction or light leading, the visibility and depth of the space are wide enough.

透过屋内各个细节中隐而不显的荷兰风格派设计相互呼应，借以营造温润大气却不失华美的居住环境。

带着荷兰风格派的几何美学，却又不显冰冷，为此中央的沙发区扮演了不可或缺的关键角色，不但可以别于一般沙发的单向坐法，与此同时又可多向度的坐卧。然而多功能的沙发并不复杂，复杂的是人们坐在沙发时所面对的屋内风光，透过面对而决定了沙发的属性。不需大刀阔斧的建设，不需轻盈的指引，空间的可见性与深度就已经足够敞开。

260 | 亚太名家荟萃·席卷港台

李文心 / Wenxin Lee
设计总监

许天贵 / Tiangui Xu
主持建筑师/设计总监

传十室内设计有限公司系由许天贵建筑师与李文心设计总监创办领导，以建筑专业养成背景为服务基础。对于风格并无特定之框限，系顺应客户之偏好为起点，重视环境脉络，擅长调和机能、美学、设备、家饰等元素，追寻独特的设计氛围与美学品味，为业主提供跳脱传统框限的设计格局。

2016年获德国IF DESIGN AWARD居住空间奖，作品"灰色之境"。在2015年亚太室内设计大奖APIDA 居住空间类 TOP10，以及2015年中国金堂奖居住空间优秀设计奖和2015年国家金点设计奖Golden Pin Design Award 空间设计类入围。而且2007~2013年连续7年获得"台湾室内设计大奖"（TID Award）住宅空间类TID奖，累计共10件作品获奖，3件作品入围。

大景 轴线
THE VENUE WITH A VIEW, A TALE OF TWO AXES

设计公司｜传十设计
设 计 师｜许天贵、李文心
项目地点｜中国台湾
项目面积｜240 m²
主要材料｜木质板、乳胶漆、布艺等

DESIGN CONCEPT ｜ 设计理念

This project embodies panoramic views of the river and the cityscape. As a 'smart home', the client utilizes it as a vacation and entertainment venue for an intelligent and healthy lifestyle.

The entire layout is modified to enhance the advantageous view of the landscape and to establish a spatial order. The cross axes has one axis from the entrance that connects the public and private spaces. It also has another axis that starts with the kitchen, dining room and living room and ends with the focal point, the floor-to-ceiling window view.

　　本案拥有壮阔河景与都会景观之大宅，屋主将作为度假与宴客的会所，同时能兼具智慧、健康享乐的"智慧住宅"。

　　整个住宅格局的改造，是以发挥景观条件优势与建立空间秩序为主。两条十字正交之轴线，其一的入口轴线，串连公私领域空间，另一景观轴线则将客厅餐厅厨房等公领域空间整合在同一线，并与落地窗景连成一气。

This project uses the smart home system to control all lighting, HVAC, curtains, media and stereo systems. Most materials used are green certified products, for promoting a better and healthier lifestyle. To reduce the cold ambience of the stereo and electronic systems, earthy tones and materials are used, such as wood veneer, travertine and stone veneer floorings. Through the gradation of light and dark materials, the spaces are each given a different expression.

本案大量运用科技控制设备，可控制灯光、空调、窗帘、影音器材等设备，在材质方面则大量采用具有绿建材认证之材质，提升便利生活与健康。为了降低影音设备等科技产品的冷冽感，特别采用大地色系材质，如木皮、洞石、仿岩面地砖等材质，营造低调时尚的温润氛围。在深浅不一的材质运用下，让各空间都有不同的表情。

The living room has the beige travertine as the TV wall. The sofas are backed by light wood veneer walls, which continue to the dining space. The dining space is complemented by a large stained glass door which exudes artistic ambience. The wine rack system nearby uses wood veneer, stones and intermediary lighting to create an underground wine cellar atmosphere. The custom-made pendant light above the dining table is likened to a wine glass runway and becomes the focal point of this space.

客厅区以米黄色洞石作为电视墙，而沙发背墙则选用浅色木皮材质并延伸至餐厅空间。餐厅在大型彩绘玻璃门的衬托下，展现独有艺术气息，一旁的酒柜空间，利用木皮、石材与间接照明设计，营造出宛如酒窖般的氛围。餐桌上方的订制吊灯，是以酒杯的走秀平台为概念，成为餐厅区的视觉亮点。

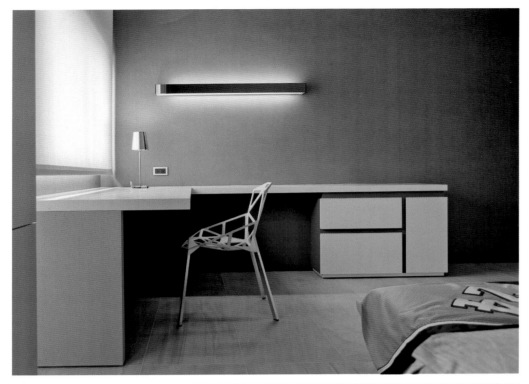

The entrance to the private quarters uses a concealed door design. Behind it, the corridor is lined by grooves of recessed lighting, creating rich visual effects. The master bedroom also has a large panoramic window with a great view. The layout intentionally avoids a large walk-in closet design to enhance visual openness. The form and materials are minimal and low-key, so as not to overshadow the beautiful scenery outside. The children's bedroom has a white theme, with only added colors of yellow and purple. These are the colors of the children's favorite basketball team. The room is full of youthful vigor.

在通往卧室区域的入口，采用隐藏门的设计，门后走廊两侧是以矩形凹凸木皮墙面与错落的间接光源，展现丰富的视觉效果。主卧房同样拥有大片窗景优势，在格局设计上，刻意舍弃大型更衣室的设计，让视线更为开阔，造型与材质运用也尽量简洁低调，让窗外美景成为主角。孩房则是白色为基调，加上使用者喜爱的美国职业篮球队的黄、紫色系，让空间充满年轻活力。

唐忠汉/TT

近境制作 设计总监

唐忠汉，2003年成立近境制作，擅长现代时尚风、简约机能风、自然原始风。坚持设计源自于对生活的热情，活力、单纯、亚洲的设计理念。在我们的系列作品里，透露着强烈的地域色彩，透过完整的室内建筑概念，以材质乘载情绪，以光影纪录时间，生于亚洲，源自东方，以最真诚的人文精神，诉说着空间的故事。

2016年荣获意大利A' DESIGN AWARD，荣获韩国K-Design Award。2015年荣获INSIDE AWARD, World Festival of Interiors 住宅空间类大奖以及荣获 Design For Asia Award亚洲最具影响力设计奖和中国Idea Stone IDS年度风云人物。2014年荣获意大利A' DESIGN AWARD 银奖，还有全球先锋设计（中国）大奖Global Pioneer Design (China) Award 银奖和铜奖，以及荣获成功设计大赛Successful Design Award和TID台湾室内设计大奖，同时荣获点石奖2014中国国际绿色建筑装饰设计精英赛金奖，还有荣获 PChouse中国十大高端室内设计TOP 10，第十二届现代装饰国际传媒奖年度家居空间大奖，金堂奖中国室内设计年度评选年度十佳住宅公寓设计等。

质域
TEXTURES

项目名称｜元大之星
设计公司｜近境制作
设 计 师｜唐忠汉
项目地点｜中国台湾
项目面积｜175 m²
摄 影 师｜岑修贤摄影工作室MW PHOTO INC
主要材料｜石材、镀钛、木皮、镜面、铁件等

DESIGN CONCEPT ｜ 设计理念

The vocabulary of textures is what defines structures in a space. Contrasting structures are stacked together, blending steadfastness and a vigorous rhythm, while lines pass through the structures to balance the whole space so that it doesn't feel too heavy.

Two spatial structures are placed in to naturally create an open space. Lines of material and light serve to separate areas from each other, splitting the original area into three new sections and forming another unique section within itself. The space is split in two. One side employs a single material to give the structure depth and spaciousness, while the other side employs different materials that gradually blend into each other, so that the structures are less obvious and the line between spaces is less clear, to the point that it is nothing more than a metaphor. Lighting is scattered around the room, making the essence of the space seem flowing instead of static. The main space uses base conditions in its planning, and lets light into the room through a vertical flow. In a continuity of space, the light is scattered around the space through multi-level textures. The light and shadows, thus reflected, create a fun feeling throughout the whole space.

材质的语汇界定空间的量体，对比量体错层堆栈，沉稳与律动相互交融，线条贯穿延伸平衡空间重量。

置入两空间量体使开放场域自然而成，以材质线条与灯光分界空间量体，增添原本一分为三的空间场域，形成另一独特区域。空间一分为二，一边以单一素材建筑手法延伸空间深度，另一边以材质渐变方式减化量体的存在感；暗喻空间的界定，错落的光跟影让空间本质交错流动。主空间在规划上运用基地条件，采用垂直动线，引光入室。延续性空间，灯光透过材质错层散布在空间之中，反射的光跟影，营造空间趣味性。

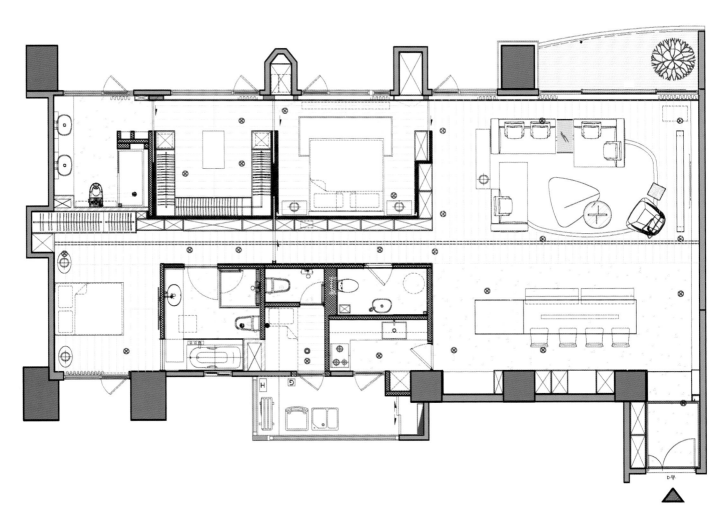

The space is further decorated by stone and wooden textures, with titanium coated over quantized space. Through reflective materials, the textures of the material themselves are continued; and through reflections and split layers, the versatile space signifies the ever-changing quality of life.

空间以石、木润色、镀钛包覆量化空间，透过反射性材料，延续材料本身的质。因反射交迭，传递空间多变的生活面向。

蔡馥韩 / Fu Han Tsai

天境空间设计有限公司 创始人、设计总监

蔡馥韩，2001年毕业于大叶大学空间设计系，2005年成立天境空间设计有限公司。从业多年，多次获奖，其中有：2011年度TOP DESIGN AWARD、2013年度Dulux空间色彩大赏—金、2014~2015年度设计影响中国十佳优秀设计师、2016 英国London Design伦敦设计大奖、2016 义大利A'Design等奖项，同时2016年度最佳空间设计创造建商四天完销纪录。

设计作品涵盖私人住宅、样板间、商业空间等，代表作品有旭阳国际精机厂办、海德一号、仁山协和、帝之苑、和唐金龙、欧洲之星等，作品收录于《台式美学》《台式哲学》《宅在台湾I》《宅在台湾II》《时尚设计盛典作品年鉴TOP100》《漂亮家居2016年百大设计师年鉴》《iw家饰88》及《亚太豪宅大赏II》等多部专业设计书刊杂志中。

精锐音悦厅
EXQUISITE AND ENJOYABLE MUSIC HALL

设计公司｜天境空间设计
设 计 师｜蔡馥韩、洪慈涵
项目地点｜中国台湾
项目面积｜165 m²
摄 影 师｜刘俊杰
主要材料｜不锈钢玫瑰金镀膜、纽约大理石材、幕光大理石、铜条、木皮、皮革、薄板石材等

DESIGN CONCEPT ｜ 设计理念

This project uses New York stone, screen light marble stone, copper, cypress sandblasted wood veneer, white saddle leather, panel stone and rose gold plating. Each material has its warm color and plain texture beauty. The designers use these elements to agitate a dialogue and aesthetics.

　　此案用了纽约石材、幕光大理石、铜条、桧木喷砂木皮、白色马鞍皮革、薄板石材、玫瑰金镀膜，每项材料皆能窥得其颜色温泽与一份无娇造之美，设计师运用了这些元素去激荡一种对话和交衬美感。

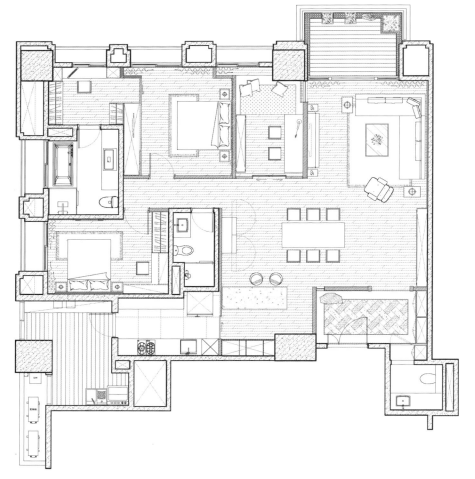

The side view wall in the porch uses fine and mellow stones, collocating with half-penetrated plating laser cutting, which presents luxurious senses. Cloth droplight matches with ink stone wall. The floor parquet uses three stones to make a gradation, which creates a layering aesthetics.

The living room background wall matches metal cooper bars with cypress sandblasted wood veneer. The design of split screen outlines bright lines. The display cabinet is based on cypress wood veneer, decorated with archaistic mirror and rocks. The stone of TV background wall creates a sedate atmosphere. The intersection with lateral flowers and straight avoidance and the inversed V-groove stone present exquisite technique, which makes the living room more magnificent.

玄关端景墙佐用调性缜致醇厚的石材，搭配半穿透镀膜雷射切割呈现奢华的感觉，布质吊灯配上墙面泼墨主石材，玄关地板拼花以三块石材去作层次表现，促成一种层次的美学。

客厅背墙以金属铜条搭配桧木喷砂木皮，分割画面的设计勾划鲜明线条，侧边展示柜桧木木皮为基底搭配仿古镜及版岩作为饰品陈列。客厅电视主墙石材营造稳重气息，以交丁方式横向对花而直向错开，石材细部局部倒V沟显现细腻的手法，使客厅空间感加大，更有气势。

The sliding door of the study continues the plating laser cutting modeling in the porch with an echo, which makes the transparent treatment having no oppressive feeling. Considering functions of the study, the bed can be used as a single guest room. Warm sandblasted wood veneer matches with rose gold plating, which creates a fashionable and leisure interior aesthetics. The furniture design foils the background and easily matches with the space, which can agitate a warm and exquisite atmosphere and create a comfortable communication corner.

The red wine cabinet in the dining room collects red wine cups, coffee cups, plates and household appliances. It is like a suitcase when closed. The theme is five-star Mini bar concept, which makes the owner bathes in a life scene and leisure tone at home and enjoy the leisure time with a cup of wine.

书房的拉门延续玄关的镀膜雷射切割造型，空间实感呼应，让空间通透的手法处理没有压迫感，考量书房要多功能的运用，卧榻兼单人床客房的使用概念，温润的喷砂木皮搭配玫瑰金镀膜，摹绘一种时尚而兼具休闲格调之室内美学，搭配设计家具为背景称托出来，家具易于和空间配衬并激荡出一种温馨及精致的氛围，同时营造出舒适的谈话隅角。

餐厅红酒柜里头收纳红酒杯、咖啡杯、盘子、家电，关起来变成行李箱，以Mini bar五星级的概念为主，让屋主回到家沐浴在一份生活感与休闲步调中，亦能啜饮水酒享受这段悠闲时光。

The 165 square meters space makes the living room and dining room wide. The sliding doors of kitchen separate cooking area from snack bar. The headboard of the guest room spreads over into the ceiling, which makes the space not dull. Considering loading weight, the bookcase of the study is deliberately designed with a 45 degrees angle, which satisfies visual beauty. The wood cabinet uses plates to present fine closing. Texture accessories are used in the space, which presents lively feelings.

The main wall of master bedroom is decorated with frames, leathers and stitches, which outlines superior delicacy. Classical collage and mirror droplights on both sides of the bed add fashionable flavors. Whole steel wall with wood basement at the end of the bed, coherent partition design and hidden sliding door of the dressing room set natural connotations of the space.

165平方米的空间让客厅餐厅书房开阔，厨房拉门隔绝快炒区及吧台轻食区，客房床头线板造型延伸到天花板，让空间不沉闷。书房中书柜层板考虑到承重的问题，刻意做了45度斜角的设计，同时也满足了视觉美观，木作柜体以线板的概念呈现细腻的收边，另外挑选有质感的配件来搭配空间，突显活泼的感觉。

主卧室主墙以画框概念绷皮革与车缝线勾勒出俨如高级订致服的精致感，加上古典设计感去拼贴，两侧床头再以镜面吊灯去做搭配，增添时尚的韵味，床尾整面钢刷处理的木皮作为基底及连贯性的分割设计，并将进入更衣室拉门隐藏，奠定了空间的自然底蕴。

江欣宜 / IDAN CHIANG

缤纷设计创办人　天空艺术执行董事

江欣宜，缤纷设计创办人、天空艺术执行董事，同时身兼亚太酒店设计协会理事、中国室内装饰协会陈设委常务副秘书长、中央美术学院艺术设计专业陈设工作室课题教授等要职。从小接受博物馆美学艺术薰陶，拥有社会学专业背景，同时亦是艺术收藏家，擅长倾听人心，关心任何细节，跳脱设计师既有的框架，展现创新想法与独特品味。提倡发散式美学哲思，认为室内设计必须找出一个亮点，从此亮点发展延伸，并将艺术结合于设计空间中，更重视空间陈设艺术氛围与人的"关系"联结及互动，营造舒适自在的奢华空间。

IDAN荣获无数国际奖项与媒体报导肯定，如台湾TID，香港APIDA，中国CIDA晶麒麟奖与美国室内设计中国室内设计年度12大封面人物等，亦将设计领域跨足艺术策展及商业策展，透过不同的呈现方式及系统整合，将会所设计及陈设美学做更完整的诠释。

享·品味
ENJOYMENT AND TASTE

设计公司 | 缤纷设计
设 计 师 | 江欣宜
项目地点 | 中国台湾
项目面积 | 231 m²
摄 影 师 | 吴启民
主要材料 | 石材、钛金属、实木等

DESIGN CONCEPT | 设计理念

The design team is good at recombining and harmoniously placing luxurious and simple texture to make the tall space full of exquisite soul and present experienced, knowledgeable and elegant life attitude of the residents in the earth tone and furnishing arrangement. The integration of stone, titanium metal and solid wood creates a hard and soft balance, leads modern luxurious and dignified charms and guides classic spirits of American residence and majestic demeanors. The furnishings use wood floors indoor and wooden grille outdoor. Plants in different shapes and sizes bring green into the space, making the residents feel the warmth of nature.

　　设计团队擅长将奢华、简朴的质感重新融和且和谐并置，使得挑高的空间中，充满细腻灵魂，于大地彩度与陈设设计的安排下，凸显居住者游历丰富、见多识广、风雅气势的生活态度。透过石材、钛金属、实木的媒材串接，造就出一柔一刚的平衡表现，更引申现代奢华的轩昂魅力，引领美式住宅的经典精神及运筹帷幄的王者风范。陈设上，延伸空间中的木地板及户外的木格栅，透过不同形式或大小的植栽，将绿色植物引入空间，让居住者感受到大自然的暖意。

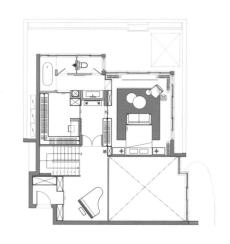

傅琼慧 / Irene Fu

由里室内设计 设计总监

2016年是由里踏入设计领域第二十二年。第一个十年，由里专注室内设计服务；第二个十年，由里除了室内设计并延伸到私人别墅及特色建筑及由内而外的生活空间；展望第三个十年，由里将整合生活建筑及室内设计施工等专业服务项目，成为全方位建筑外型空间设计整合服务团队。

由里认为完美的空间必须融合生活建筑与空间设计，美好的作品应该随着人、事、时、地、物表现不同的表情与词汇，而不是一味追求流行，更不是拷贝。称职的设计师应该是敏锐而体贴的作者，除了能提供多元的美学风格，更能完全展现空间的理性与感性。更重要的是，能满足业主实际生活需求，提供业主更多元空间的使用应用建议。

客制化奢华风 内敛气质别墅
CUSTOMIZED LUXURIOUS STYLE, RESTRAINED CHARMING VILLA

项目名称｜富丰B户
设计公司｜由里室内设计
设 计 师｜傅琼慧
项目地点｜中国台湾
项目面积｜480 m²
摄 影 师｜张家瑞
主要材料｜钢琴烤漆、复古镜、铁件、大理石、绷布、皮革等

DESIGN CONCEPT ｜ 设计理念

The designer injects this project with low-key and luxurious elements, strings home relations in the five floors transparent villa and perfectly deduces the charm and texture of the space. Entering from the garage, the reception area presents open view. Arc ceiling modeling echoes with the main TV wall, furniture modeling and spiral stair, creating a complete effect of heaven and earth.

设计师为本案注入低调奢华的元素，在五层楼的透天别墅中串起家的感动，将空间的气韵与质感完美演绎。从车库进门，迎宾区呈现开阔的意象，圆弧的天花线板呼应电视主墙、家具造型及旋转梯，形塑天地圆满的寓意效果。

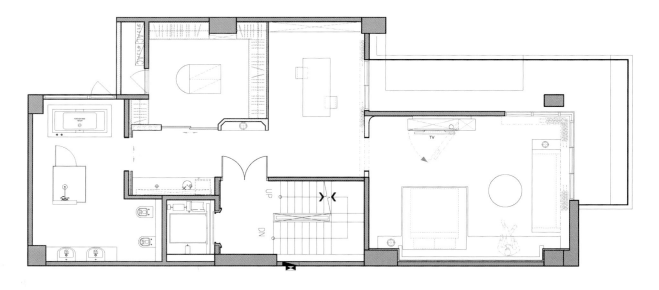

On space planning, the same visual axis connects living room, bar and dining room, which makes the kinetonema fluent and increases life interactions. With exquisite design perceptions, the textures and stereo visual modeling manifest characteristics of every field and endow every space with unique style and atmosphere. The whole space uses design method with most unique personality to present attractive home charms between modern fashion and classic aesthetics by an elegant tone.

公共空间的规划上，以同一条视觉轴线连贯客厅、吧台与餐厅，让动线流畅，同时增添彼此的生活互动，并融入细腻的设计感知，擅用质材与立体的视觉造型，将各部场域的特质发挥淋漓，也让每层空间独有风格与氛围。整体以最独特个性的设计方式，在现代时尚与古典美韵之间，以优雅的格调展现诱人的家的魅力。

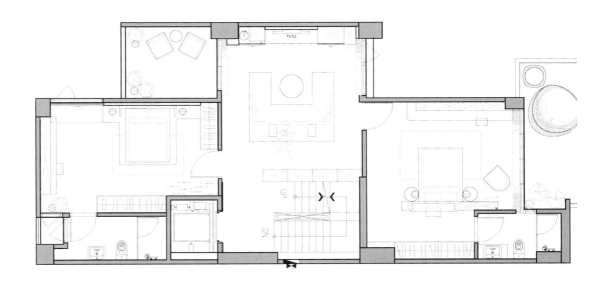

黄国桓 / Guohuan Huang

瓦第设计 设计总监

　　毕业于中原大学室内设计系，美国 Kansas State University 室内设计硕士。除接受专业正统的设计教育训练之外，更曾任职于知名设计工作室与地产开发公司，透过超过20年以上的专业训练与实务操作经验，并将个人对于人文艺术与历史文化长期关注与修养，全心投入于所爱好的设计行业中。在理性设计发想上关注使用者与环境的联结，使用者与空间尺度关系，并恪遵现代美学的主义教条；在感性上试图寻找出每个空间独特的调性与灵魂，期许引导出使用者与空间设计，色彩材料等使用的浪漫对话。

　　瓦第设计长期专注于居住行为相关的空间型态的规划设计，从地产开发商集合住宅的单元平面空间，小区公共设施与景观设计，广告企划公司之销售会馆样品屋实品屋等规划执行进而到私人住宅设计家具细部施工，均秉持着全方位专业完善的Total Solution操作执行模式。我们专注仔细地凝听业主真实的需求与声音，有系统有效率的分析客户需求与预算执行，期许能够提供客户完整且美好的设计服务与施工质量。

优享透心美宅
ENJOYING THE BEAUTIFUL RESIDENCE

项目名称 | 大观自若
设计公司 | 瓦第设计
设 计 师 | 黄国桓
项目地点 | 中国台湾
项目面积 | 270 m²
摄 影 师 | 墨田工作室
主要材料 | 木皮板、天然大理石、铁件等

DESIGN CONCEPT | 设计理念

This superior residential building is in the 14th floor of 26 floors, located in the science and technology city Hsinchu where the economic activities are highly active. The front of the building has a 25 meters daylighting opening above and a 15 meters balcony. There is a river which windingly passes through the twin cities of science and technology in Taiwan economic area. Across the river, it faces with the high-rise residential buildings.

这是一间总高26层位于第14层的高级住宅大楼，位于经济活动高度活跃的科技城新竹，建物大楼正面拥有超过25米以上的采光开口与15平方米的大阳台，前方是一条蜿蜒穿越整个台湾经济橱窗科技双子城的河流，跨过河便与对岸的高层住宅建筑群相望。

The left is continuous Central Mountain Range towering above the ground in Taiwan which is 3000 meters high while the right is endless and vast Taiwan Strait. The middle is transportation bridges such as expressway, thruway, railway train and light rail which stretch across the streams. Everyday the sun rises slowly from the mount top in the morning and goes down into the ocean in the evening on the left side. In spring the new grasses of the sports garden on both side make the front scenery green. In summer the heat sunshine makes the distant sea surface shine flowing silver. In autumn all the Miscanthus floridulus in the river become white and dance with the wind. In winter when the cold current comes, if luckily you can see rare snow of the subtropical climate region in the chine line of the Central Mountain Range. From early morning till late light, all kinds of transportation shuttle busily on the bridges and roads. And the opposite residential building becomes higher and higher above the ground.

左侧是拔地而起绵延不断高耸达3000米以上台湾屋脊中央山脉，右边远望则是一望无际的台湾海峡，中间夹杂了横跨溪流的高速公路、快速公路、火车铁道，轻轨电车等交通工具的多座桥梁。每天左侧会迎接清晨阳光从山顶慢慢升起，傍晚的太阳则是沿着溪水缓缓落入海洋。春天的时候两岸河川运动公园的新生草坪让前方的景色充满了嫩绿；夏天炙热的阳光让远方的海闪耀着跳动的银光；秋天在河道中间所有的五节芒都白了头随着季风摇曳摆动跳舞；冬天如果寒流来时，运气好的话，还能在中央山脉棱线上看见在亚热带气候区少见的积雪。从清晨到深夜各式不同的交通工具忙碌地穿梭在桥梁与马路上，对岸的住宅大楼则是一层一层的从平地而起。

Facing with the priceless landscape residence with a coexisting development of vast natural geography and human material civilization, the designer presents the authentic and moderate superior exterior view, which makes the interior residents deeply feel the seasonal natural landscape changes and enjoy the profound connotation and high availability of a long-term development of human civilization on space planning.

面对这个广阔的自然地理与人类物质文明发展并存的无价景观住宅，设计师的主轴便是将这个室外优越视野忠实谦虚地呈现出来，让居住在室内的使用者能够深刻地感受一年四季大自然的景观变化，同时在室内空间规划上享受到人类文明长期发展的深刻内涵与高度便利。

The designer keeps maximum lighting area of the original windows and avoids higher furniture and objects. So he chooses suitable scale furniture as the design character of the space. Opening the door, there is an impressive view. The designer wants to give an amazing vision to visitors about the tone of the porch. So he chooses contemporary Bauhaus Poltrona frau bench, collocating with cloth collage in different colors. The fashionable droplight echoes with modern artistic style of the hostess. According to knowledgeable humanistic quality of the host, the designer chooses Turkey Vintage carpet in low chroma, collocating with leather texture metal screen, primitive carpet, fashionable bench, white piano, painting door and black texture marble floor. The designer makes full use of conflict contrast of materials and colors to make the porch present personal aesthetics and characteristics of the owners.

设计师将原有开窗的部分尽可能保留最大的采光面积，避免过高的家具或对象阻碍了视野，因此挑选了尺度适宜的家具作为空间设计主角。一开门就面对令人印象深刻的视野，设计师也希望让访客对于玄关的调性同样有眼睛一亮的感觉，所以挑了造型有当代包浩斯设计感的Poltrona frau bench，搭配不同颜色的布品拼贴，再挂上时尚的吊灯来呼应女主人的现代艺术风格。依照男主人博学的人文知识素养，则是挑选了土耳其的

低彩度Vintage carpet，搭配皮革纹理的金属屏风、古朴的地毯、时尚的bench、白色素面钢琴烤漆的门板及充满黑色纹理的大理石地板，设计师充分利用材料与色彩的冲突对比，让这个小玄关空间瞬间将屋主的个人美感与特质崭露无遗。

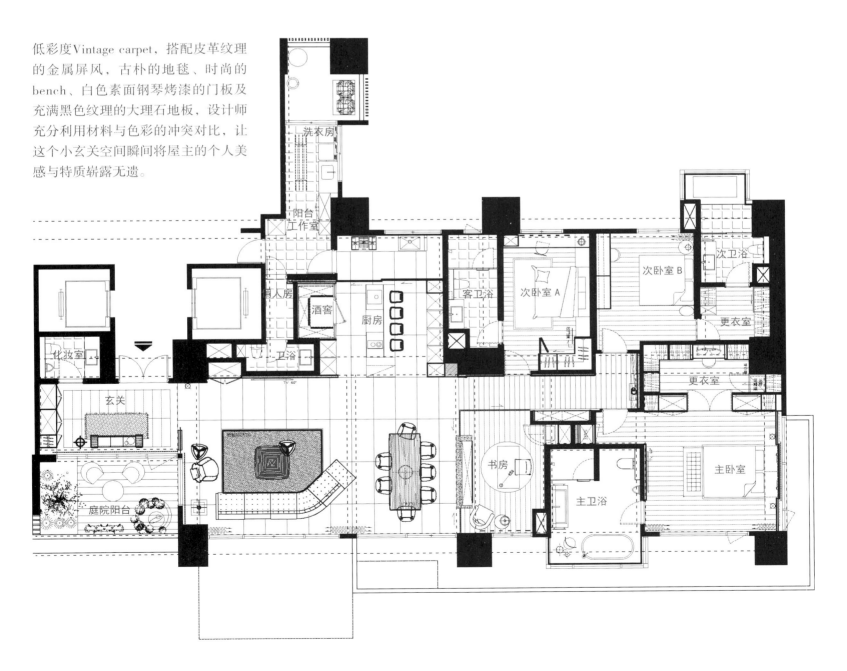

Considering that the liivng room, dining room and kitchen are the main living spaces of the owner, so the designer integrates them into one big space to make the design. In order to avoid higher furniture which may resist the lighting and the view and complicated modeling and strong colors which may mix up outside green scenery, so the designer chooses low and concise furniture as sapce character.

The designer thinks that the dining room and kitchen are the most frequently used spaces in the house, so he creates an independent wine cellar to make it convenient for the family and friends to enjoy when eating at the island and the dining table. In order to echo with the Taiwanese cooking habits, a transparent hidden kitchen sliding door is set to keep away the annoying smell from the outside space when cooking. The designer adds some philosophy into the design of dining room. The background adopts the same stone as in the porch floor. The collage technique creates the effect of star sky. Spain VIBIA COSMO droplight hangs ups and downs in different sizes, collocating with walnut table with plant fossils inside. The Japanese iron pot and plain pottery represent land on the earth and people. It hopes that the dining room can present harmonious and endless concepts of the universe.

考虑客厅、餐厅与厨房是业主平日的主要生活空间，设计师将此三个空间视为一个大空间处理。同时避免选择尺度过高过大的家具背阻碍了采光与视野，或是过于繁复的造型与强烈的色彩表现混淆了户外的绿色景观，因此设计师选择了摆设低矮简约的家具作为空间家具主角。

餐厅厨房是设计师认为家人使用最频繁的空间，因此在厨房创造出有一个独立的藏酒间，方便家人与朋友在中岛与餐厅用餐时共享。而为了呼应台湾人的烹调习惯，厨房设置了隐藏透明拉门，使食物在烹煮时能够阻绝恼人的味道至厨房以外的空间。餐厅则是设计师偷偷加入了一些个人哲理的空间，餐厅背墙选用了跟玄关地坪相同具有丰富纹理的黑色大理石材，利用拼贴手法创造出星际银河般的线条效果，西班牙VIBIA的COSMO吊灯，高低起伏大小不一的悬挂着，搭配上胡桃木内崁植物化石的餐桌。餐桌上放置着日本铁壶与素烧陶花器，象征着地球的土地与生活上面的人们，希望这个餐厅空间呈现宇宙和谐、生生不息的概念。

图书在版编目（CIP）数据

亚太名家荟萃：豪宅赏析 / 深圳视界文化传播有限公司编． -- 北京：中国林业出版社，2017.3
ISBN 978-7-5038-8916-5

Ⅰ．①亚… Ⅱ．①深… Ⅲ．①室内装饰设计－作品集－亚太地区－现代 Ⅳ．① TU238.2

中国版本图书馆CIP数据核字（2017）第 023445 号

编委会成员名单
策划制作：深圳视界文化传播有限公司（www.dvip-sz.com）
总 策 划：万绍东
编　　辑：杨珍琼
装帧设计：潘如清
联系电话：0755-82834960

中国林业出版社 · 建筑分社
策　　划：纪　亮
责任编辑：纪　亮　王思源

出版：中国林业出版社
（100009 北京西城区德内大街刘海胡同 7 号）
http://lycb.forestry.gov.cn/
电话：（010）8314 3518
发行：中国林业出版社
印刷：深圳市雅仕达印务有限公司
版次：2017 年 3 月第 1 版
印次：2017 年 3 月第 1 次
开本：235mm×335mm，1/16
印张：20
字数：300 千字
定价：398.00 元 (USD 79.00)